Beyond Standardization: Accounting for Gripping System Effects to Improve Geosynthetic Interface Characterization

Richard

TABLE OF CONTENTS

1 INTRODUCTION

1.1 Background

Geosynthetics, according to ASTM D4439 (2018), are planar products manufactured from synthetic polymeric materials and can be used with soil, rock, earth, or other geotechnical engineering related material as an integral part of a man-made project, structure or system. They are a product manufactured in rolls that are classified into different groups based on the method of manufacturing and intended use. These groups include geotextiles, geomembranes, geosynthetic clay liners, geogrids, geonets, geofoam etc.

Geosynthetics are currently employed in most geotechnical-related projects like landfills, road construction, steep slope stabilization, water management and basal reinforcement of the foundations. A typical example of geosynthetics application is shown in Figure 1-1.

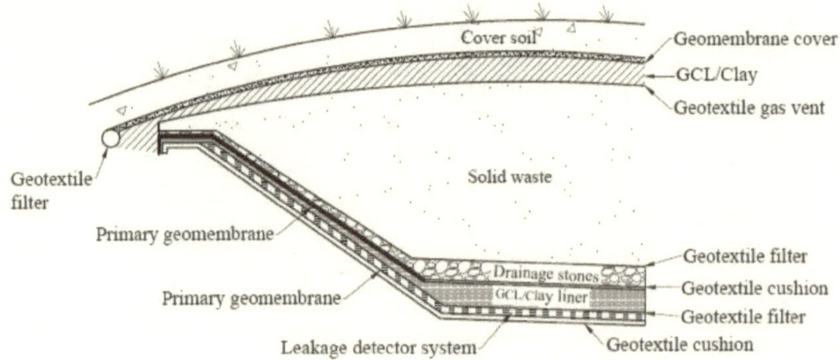

Figure 1-1: Example of landfill liner consisting geosynthetics (after Geosynthetic World, 2018)

The application of geosynthetics in geotechnical engineering projects has rapidly increased since the late 1950s, when they were introduced on the market (Textile Centre of Excellence, 2018; Geosynthetic Materials Association, 2016). This has resulted in significantly improved performance, cost reduction and environmentally friendly of geotechnical structures as compared to the utilization of conventional materials i.e. Clay. It is important, however, to note that geosynthetics are mostly not individually installed on site, but work in combinations of each other. This results in the introduction of many interface plans on/in a structure where movement or even slippage could occur when loaded, thus, causing failure of the structure (Kalumba 2018).

One of the critical factors in preventing failures of structures that consist geosynthetics is the use of appropriately characterized geosynthetic interface shear friction in the theoretical design of the structure. The interface shear friction parameters such as the frictional angle and adhesion must, therefore, first be determined.

The interface friction angle and adhesion are measured in the laboratory mainly using the direct shear apparatus in accordance with ASTM-D5321 and ASTM-D6243 standards (Sikwanda et al. 2018). Although these laboratory tests are standardized, the quality of the results can be largely affected by several factors such as specimen gripping systems, shearing rate, applied normal stress, and the type of geosynthetic test specimens (Fox & Stark 2004). Among these factors, the former is considered to be a major source of dissimilarities in the results (Fox & Kim 2008). This is because when the geosynthetic being tested is not sufficiently secured to the shearing blocks by the gripping system, it experiences progressive failure and shear strength that deviates from the actual field performance (Fox et al. 2004). This could lead to unsafe, cost ineffective, etc. design of projects with the respective geosynthetic materials. This research, therefore, was undertaken to investigate the effects of the specimen gripping system on shear strength at the geosynthetic/geosynthetic interface using a direct shear device.

1.2   Research Justification

The need to obtain representative interface shear strength parameters of materials, in many geotechnical engineering projects is key in designing optimum structures. This is because the correct characterization of mechanical properties is useful in preventing failures in structures like landfills, retaining walls, steep-sided embankments and cuttings etc.

To date, the use of geosynthetics has expanded rapidly nearly in all geotechnical related projects. It is, however, imperative to note that geosynthetics are mostly not individually employed, but as a combination with each other. This is typically common in applications like landfill liners, where several layers of geosynthetics are installed prior to waste placement to improve the stability of the in-situ soil and prevent leakage of contaminants to the environment (Buthelezi 2017). An important characteristic of the liner systems with respect to slope stability is the interfacial shear resistance manifested between two geosynthetics or geosynthetics and the adjacent material i.e. soil. This is in line with the majority of failures reported in the literature (i.e. Jones & Dixon, 1998; Mitchell et al., 1993) which were controlled by slippage at the

interfaces. Therefore, it is important to define the right parameters in the theoretical interface shear behaviour analysis of structures that involve geosynthetics (i.e. landfills). This study, however, concentrated solely on the interface characteristics between geosynthetics as they are mainly considered the most critical (Visser 2018).

An international code of practice for determining the interface shear strength of geosynthetic/geosynthetic and soil/geosynthetics, ASTM D5321/6243, specifies that a direct shear apparatus is to be used to determine the geosynthetic/geosynthetic interface shear parameters for various normal stresses for design. This equipment is typically designed to shear geosynthetic specimens between a movable shearing block and a static reaction block, each covered with an abrasive gripping surface. This surface transfers the applied normal load to the specimen to ensure uniform shear strain at failure (ASTM D6243 2018). Currently, laboratories use a variety of gripping surfaces in geosynthetic/geosynthetic interface shear testing and the preferred surface remains a topic of concern. This has led to differences in many results reported and consequently, difficulties in their interpretation. Moreover, the available gripping surfaces are at times reported to not sufficiently secure the tested specimens to the shearing plates (Fox et al. 2004; Allen & Fox 2007).

The most commonly employed gripping surface includes a nail plate, sandpaper, glue/adhesive bond, and metal textured surface, which are mostly used with a clamping device to form a gripping system. The clamping device provides extra confinement to ensure failure takes place on the desired interface (ASTM D5321 2017; ASTM D6243 2018). Great effort has been made by researchers (i.e. Fox et al. 1997) to compare the effects of different gripping systems used in geosynthetic interface shear testing, but to date, there has been no investigation conducted to evaluate the effects of the nail plate and sandpaper-gripping system. For this reason, this study was conducted.

1.3 Research Objective

The aim of the study was to investigate the effects of two commonly used gripping systems on shear strength at the geosynthetic/geosynthetic interface, using a direct shear device. Specifically, the research utilised the nail plate and the sandpaper. It was anticipated that the study would outline some clarification about the competency of one gripping method over the other for investigating the interface shear behaviour between two geosynthetics. The specific objectives

of this study were to assess and compare the effects of the nail plate and sandpaper on the interface shear stress responses and interface shear stress parameters as a whole.

The reviewed literature shows that the direct shear tests at the geosynthetic/geosynthetic interface were mainly conducted using a single interface test approach. In this study, the experiments were performed not only by the single interface test approach, but also by a multi-interface test configuration. It was, therefore, desirable to compare the geosynthetic interface shear strength obtained from the single interface tests, with that from the multi-interface tests in terms of interface shear strength. By doing so, a compatibility of a single and multi-interface test in the determination of the critical interface was investigated.

1.4   Scope of the Study

This book focused on the laboratory investigation of the effects of the nail plate (NP) and sandpaper (SP) gripping system on a shear strength interface at geosynthetic/geosynthetic using a 305 mm x 305 mm direct shear device in accordance with the ASTM D 5321 and D 6243 standards. A modern landfill liner consisting of the geotextile-cushion/geomembrane, geomembrane/geosynthetic clay liner and geosynthetic clay liner/geotextile-filter interface was utilized to achieve this objective. Therefore, the results obtained in this study were limited to the conditions considered herein.

1.5   Organization of the Study

This book investigated the effects of specimen gripping systems on shear strength at the geosynthetic/geosynthetic interface. Hence, to provide the reader with the significance of the investigated problem, Chapter 1 presented the background of the study with the justification, objective and scope outlined. The relevant literature on previous studies and other areas of importance in relation to the research were provided in Chapter 2. The material properties and methodology adopted in the study were explained in Chapter 3, and Chapter 4 presented the details of the research results and their discussion. The practical application consisting of a design problem utilizing the findings of this investigation was presented in Chapter 5. Finally, in Chapter 6, the conclusions of this study were drawn based on the findings and recommendations made for further research.

2 LITERATURE REVIEW

2.1 Introduction

This section aimed to review the relevant studies which have been conducted in reference to geosynthetics interface shear strength testing with emphasis on the gripping systems used. The information was then used to identify the gaps in the existing knowledge and to define the aim and objectives of the book. A summary of the discussion is presented at the end of the section.

2.2 Overview of Geosynthetics

Geosynthetics is a generic term that consists of two parts, geo and synthetics (Oriokot 2018a). The first part, 'geo', essentially means the earth or soil. The second part, 'synthetics' refers to the man-made synthesis of chemicals, especially to imitate a natural product. These products are a rapidly developing family of geomaterials been used in a wide range of civil and/or geotechnical engineering projects due to their effectiveness in solving geotechnical related applications (Hardie 2018b; Oriokot 2018b; Stripp 2018).

2.2.1 Function of Geosynthetics

In 1998, Koerner reported that; knowing the main functional characteristic of the geosynthetics enables the use of a "design-by-function" methodology (Lopez-Anido & Naik 2000). Hence, geosynthetics are used to perform one or more of the following five key functions:

- Moisture barrier – Geosynthetics can be applied as a moisture barrier to essentially prevent or limit the migration of fluids within a structure. Typical application includes landfills, road construction and water reservoirs. Figure 2-1 shows the typical application in water reservoirs.

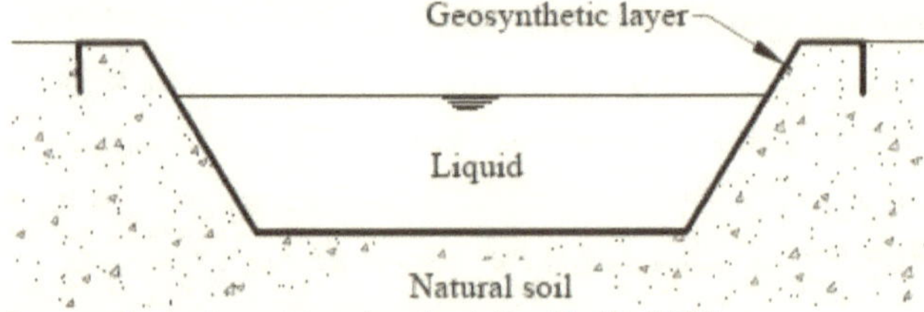

Figure 2-1: Use of geosynthetics in moisture barrier (after Shukla 2012).

- Filtration – Can be used to restrict the movement of particles subjected to hydrodynamic forces while allowing the passage of fluids with little/ no increase in pore pressure to the surrounding material. Figure 2-2 gives a schematic diagram that represents the process.

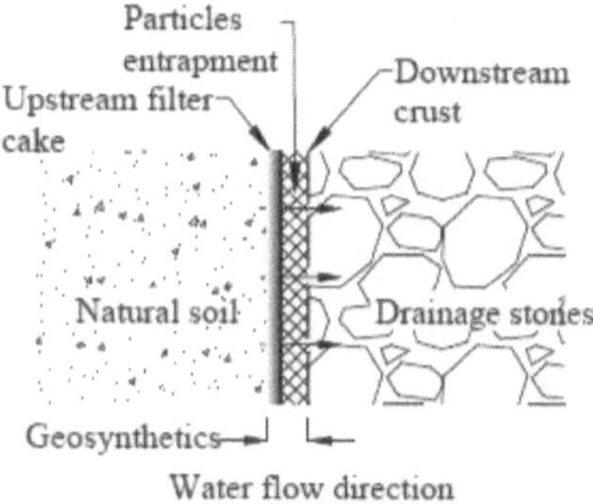

Figure 2-2: Typical application of geosynthetics in infiltration functions (after Veylon et al. 2016)

- Separation – A geosynthetic is installed between two dissimilar materials to keep the integrity and function of the structure intact by preventing the materials from intermixing. Figure 2-3 shows the typical application in road construction were geosynthetics are used to prevent road base materials from penetrating into fine underlying soft subgrade soils, thus maintaining design thickness and roadway integrity.

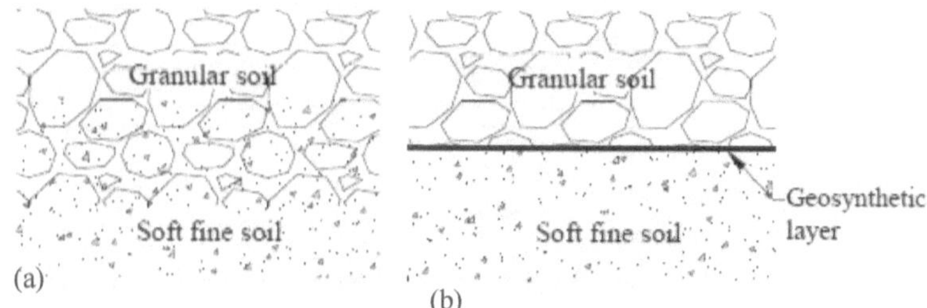

Figure 2-3: Example of geosynthetics in soil separation function; (a) without geosynthetic layer (b) with a geosynthetic layer (after Samirsinh 2016).

- Reinforcement – Geosynthetics improves the mechanical characteristics of the construction material resulting in increased structural stability. The landfills, road construction, soil slopes, bridge abutments, box culverts/bridges, and soil arches are some

of the typical applications of geosynthetics for this application. Figure 2-4 shows an example of the use of geosynthetics in slope stability.

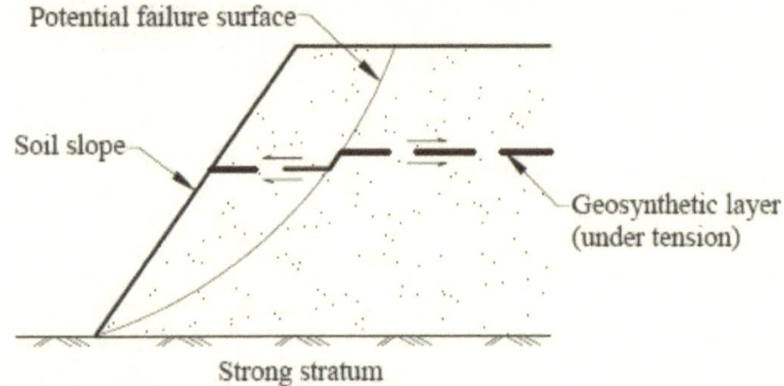

Figure 2-4: Typical application of geosynthetics in slope stability function (after Shukla 2012).

- Drainage – Geosynthetics, typically geotextiles, are used in drainage systems to collect and transport precipitation, groundwater and/or other fluids in the plane of the fabric, thus, allowing drainage of liquids away from a structure or system. Typical application includes road construction and lateral support structure. In Figure 2-5 the common application of geosynthetics in drainage is shown.

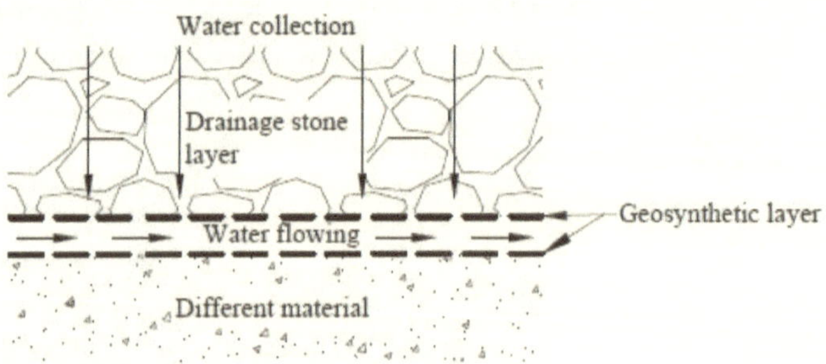

Figure 2-5: Use of geosynthetics in drainage application (after Suntech Geotextile Pvt. Ltd, 2018)

2.2.2 Geosynthetics Classification

Geosynthetics, according to the European International Organization for Standardization - EN ISO 10318 (2000), can be classified into two main categories; permeable and impermeable. These groups are further subdivided according to the methods of manufacturing i.e. geotextiles, geocomposites and geomembranes (Miuzzi 2012). Figure 2-6 briefly describes each category.

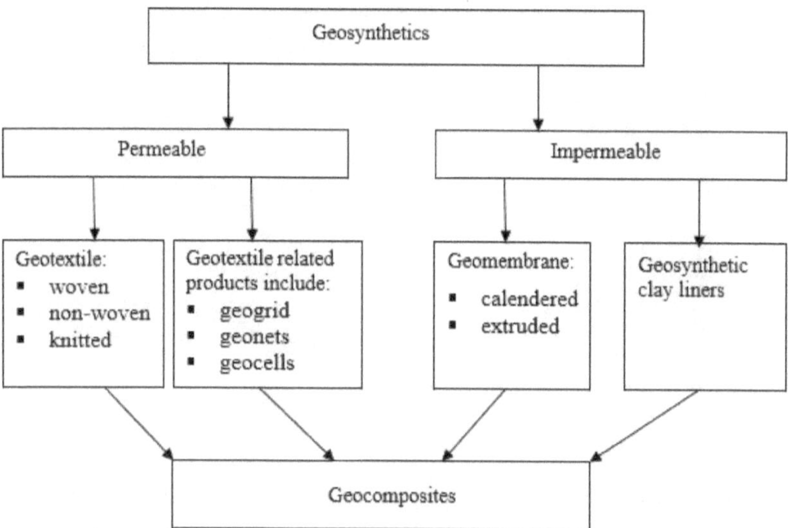

Figure 2-6: Geosynthetics classification based on EN ISO 10318 (2000) (after the Textile Centre of Excellence 2018)

This study, however, did not consider the geotextile-related products (see Figure 2-6). Therefore, the geotextile, geosynthetic clay liners and geomembranes were discussed as they were the main interest of the research work.

2.2.2.1 Geotextiles

Geotextiles are permeable geosynthetic that comprise solely of textiles and are placed within or adjacent a structure to enhance its engineered performance (ASTM D4439 2018; Textile Centre of Excellence 2018). They can perform several functions in geotechnical engineering applications that include separation, filtration, drainage, reinforcement, and protection (ASTM D4439 2018). The development of modern geotextiles materials includes woven, non-woven and knitted, as described below:

2.2.2.1.1 Woven geotextiles

Woven geotextiles are a group of geotextile materials manufactured by the uniform and regular interweaving of threads or yarns in two directions as shown in Figure 2-7. These fabrics are typically used in high strength property applications like soil separation, reinforcement, load distribution, filtration, and drainage where filtrations requirements are less critical.

Figure 2-7: Typical example of woven geotextile

2.2.2.1.2 Non-woven geotextiles

Non-woven geotextiles are light in weight products commonly used in drainage, filtration, and stabilization applications. These products are manufactured from a felt like fabric or continuous filaments that random placement of threads in a mat and bonded by heat-bonding, resin-bonding or needle punching as shown in Figure 2-8. They are typically classified into one of three categories: lightweight, medium weight or heavy weight.

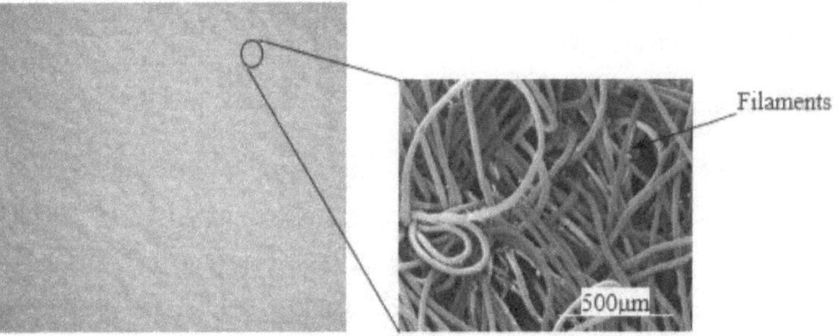

Figure 2-8: Typical example of a non-woven geotextile

2.2.2.1.3 Knitted geotextiles

Knitted geotextile are products manufactured using a knitting process in which a series of loops of yarn are interlocked to form a continuous planar structure. The yarns are kept straight and parallel to each other and aligned with the fabric's load-bearing directions as illustrated in Figure

2-9. These yarns introduced are placed in a fabric structure in four directions, warp, weft and diagonally, to give multiaxial strength.

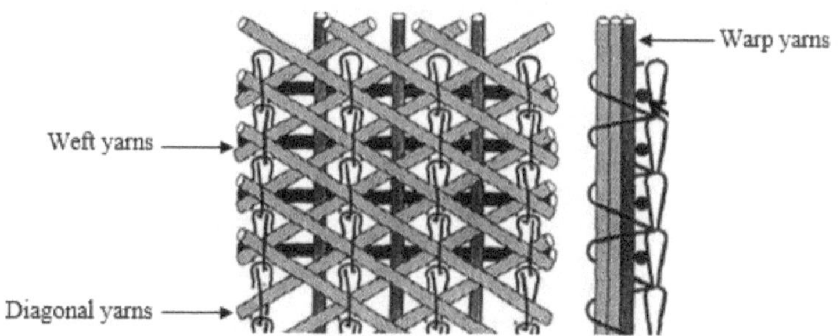

Figure 2-9: Typical warp knitted multiaxial structure (after Textile Centre of Excellence 2018)

2.2.2.2 Geomembranes

The ASTM D4439 (2018) defines geomembranes as impermeable membranes used in connection with geotechnical engineering related material as an integral part of an engineered structure designed to limit the movement of liquid or gas in a system. They are continuous flexible sheets formed either by high or low-density polyethene materials, which are combined with thermal and UV stabilizer, usually carbon black. (Textile Centre of Excellence 2018).

As reported in Figure 2-6, there are two general categories of geomembranes; calendered and extruded. Calendered are membranes manufactured by working and flattening a melted viscus formulation between counter-rotating rollers (Oriokot 2018a). The available calendered geomembranes on the market include polyvinyl chloride (PVC), chlorosulfonated polyethene (CSPE), chlorinated polyethene (CPE) and polypropylene (PP).

Extruded geomembranes are formed by melting polymer resin or chips and pushing the molten polymer through a die of the desired cross-section using a screw extruder (Oriokot 2018a). This process gives the completed product an excellent surface finish and a good workability, as the material only encounters shear and compressive stresses during the manufacturing stage (Oriokot 2018a). The High-Density Poly-Ethylene (HDPE), lower density Poly-Ethylene (LDPE), very flexible polyethene (VFPE) and polypropylene (PP) are some of the most common extruded geomembranes (Oriokot 2018a).

In geotechnical engineering, geomembranes are widely used in applications like landfills, mining and water as liquid or gas containment barriers (Bhatia & Kasturi n.d.; Bouazza et al. 2002). To the researcher's knowledge, the HDPE, is the most widely used geomembrane, to this date, for that purpose in South Africa landfills. It is herein a brief description of an HDPE.

High-Density Poly-Ethylene

HDPEs are commonly used geomembranes globally with the selections percentage been 95 in any geomembrane application project, (Scheirs 2009). The HDPEs are typically manufactured as a blown film or flat sheet product and are the cheapest (per mm thickness) industrial geomembrane commercially available, (Scheirs 2009). They have high strength, excellent mechanical and chemical resistance, hence, mostly considered as a first geomembrane material for any geomembrane application unless there are sound scientific reasons for not using them, (Scheirs 2009). However, HDPE geomembranes have a poor puncture resistance, a high degree of thermal expansion and potential for stress cracking, (WordPress 2010). The modern HDPE geomembrane includes the mono and double smooth geomembranes and mono and double textured membranes as shown in Figure 2-10. The smooth HDPE geomembranes are mainly applied in projects requiring low permeability, excellent chemical and ultraviolet resistance properties while the HDPE textured geomembranes are generally applied in high shear strength projects due to its highest multi-axial performance, (GSE Environment 2017).

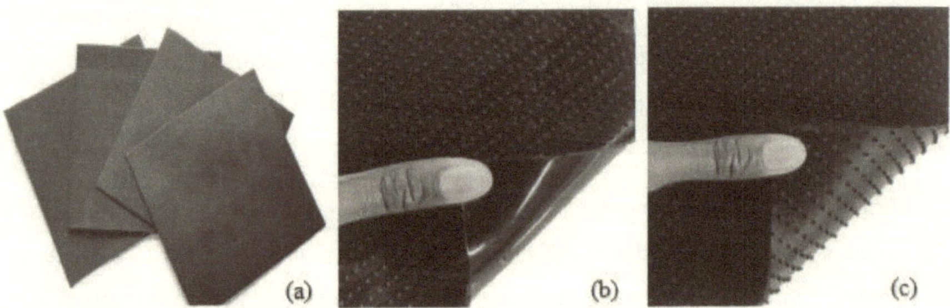

Figure 2-10: HDPE Geomembrane; (a) Double smooth, (b) Mono-Textured, (c) Double Textured

2.2.2.3 *Geosynthetic Clay Liners*

Geosynthetic Clay Liners (GCLs) are manufactured hydraulic barrier consisting of a core of bentonite clay bonded to two geotextiles and/or geomembrane materials (ASTM D4439 2018).

They are mainly used as an alternative to compacted clay liners in landfills due to the good hydraulic performance they offer at a lower-costs and easier constructability (Oriokot 2018a).

GCLs typically consist of a layer of un-hydrated, loose granular sodium bentonite, which are held together by needling, stitching and/or chemical adhesive to give the structure its internal shear resistance (Zornberg & Christopher 1999; Lopez-Anido & Naik 2000; Textile Centre of Excellence 2018). The sodium bentonite provides an effective hydraulic seal to the adjacent materials due to its high swelling capacity and low permeability. However, studies show that hydrated sodium bentonite is one of the soils with lowest shear strength, so their use, e.g. in a landfill, requires a careful shear strength assessment before they are installed (McCartney & Swan 2002). Conversely, the "carrier" geotextiles or geomembranes offer a long-lasting resistance to physical or chemical break-down in harsh elements (NILEX Civil Environmental Group 2010).

Apart from landfill applications, GCLs can be employed in the transport facilities as environmental protection barriers, especially in areas (i.e. in roads and railways) where the risk of accidental chemical spills are likely to happen. They can also be installed in underground storage tanks as secondary liners for underground protection at fuel stations (Rouncivell 2005).

In Figure 2-11, two main categories of GCL are shown, i.e. unreinforced and reinforced. Unreinforced GCLs mainly consist of a layer of bentonite clay mixed with an adhesive then attached between two geotextiles or geomembrane, backing components with additional adhesives (McCartney & Swan 2002). Reinforced GCLs on the other hand, generally consist of a layer of bentonite clay placed between geotextiles and/or geomembrane, which is bonded together by a random assortment of fibres, punched through the GCL by threaded needles (McCartney & Swan 2002).

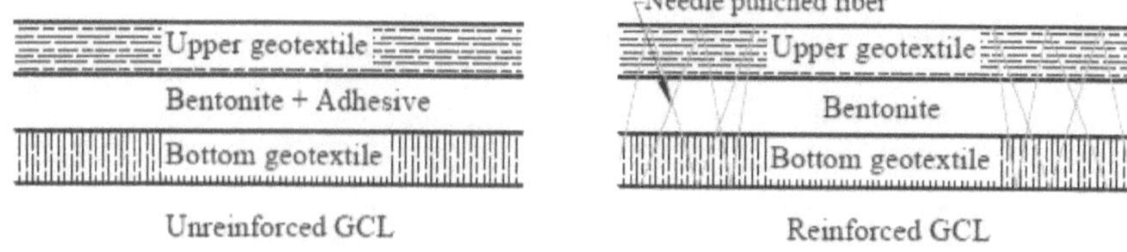

Figure 2-11: Two main categories of GCL (after McCartney & Swan 2002).

2.3 Geosynthetics in Landfills

2.3.1 Introduction

In this research, the effects of the gripping systems in interface shear strength between two or more geosynthetic layers were investigated. A modern landfill liner system was selected and described herein as an applicative example. This is because geosynthetics are popularly used in landfill liners, for that matter in South Africa.

2.3.2 Landfill

A landfill, according to the South African Department of water affairs and forestry (1998), is an environmentally acceptable facility designed for safe disposal of solid waste. It is considered to be the cheapest and most convenient method of disposing of solid waste that cannot be re-used, recycled or treated (Westlake 1995).

To date, the adoption of engineered/modern landfills as a means of disposing of solid waste (i.e. Municipal) has become a standard practice in many countries, including South Africa. The primary objective of these landfills is to maximize disposal capacity by constructing side slopes at a steep angle while isolating the contaminants such as leachate that may be present in the stored waste stream from the environmental health of humans.

In Figure 2-12, the three main engineering elements of the landfill are depicted; basal liner, side slope liner and capping liner, as categorized by the Geosynthetics Interest Group of South Africa-GIGSA (2018). These elements mainly consist of multiple geosynthetics that form a low permeable barrier termed composite liner system, which is installed in the landfill sites prior to waste placement. In this study, however, the capping liner was not considered.

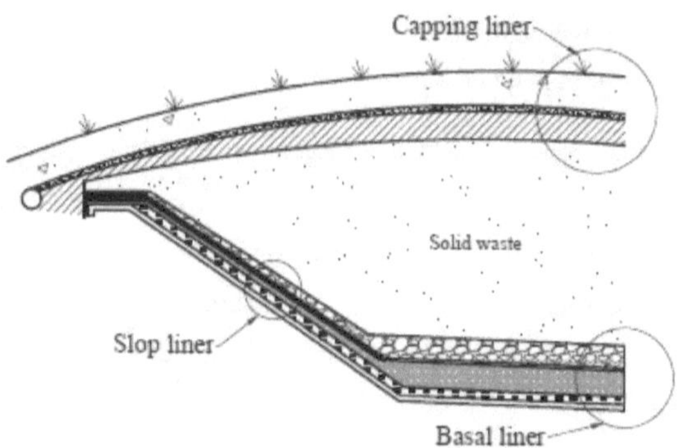

Figure 2-12: Three main engineering elements of a landfill liner (after GIGSA, 2018)

The design of the liners may vary from country to country but, according to the Department of Environmental Affairs (DEA) - South Africa (2013), the typical traditional installation combinations for a hazardous waste landfill were as shown in Figure 2-13.

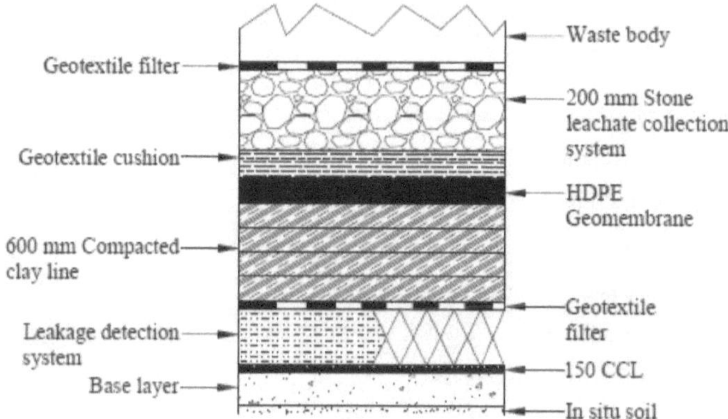

Figure 2-13: Installation layers for a tradition hazardous waste landfill (after DEA 2013).

Nowadays, the design of the landfill liner system has evolved with the increase in environmental regulations, siting hearings, and increased public awareness. One of the significant changes is the use of the GCL as an alternative to the traditionally thick compacted clay layer. The currently proposed engineered liners, consists essentially the same components, in the same order as shown in Figure 2-13, with the compacted clay layer (CCL) being replaced with the GCL. The GCL has been found to be easy and quick to install, relatively cheap, has a greater tolerance for differential settlement and a better self-healing ability as compared to clay (Oriokot 2018a). Additionally,

the GCLs have excellent hydraulic conductivity performance ranging from 2 x 10^{-10} to 2 x 10^{-8} m/s depending on the applied pressure (Rouncivell 2005).

The addition of another geosynthetic interface, however, may introduce a potential slip failure in a liner system (Kalumba 1998; Cilliers 2018c). Therefore, to ensure the performance integrity of the liner, landfills are designed based on the classes i.e. A, B, C or D particularly in South Africa (Department of Environmental Affairs (DEA) - South Africa 2013). The groups of landfills are mainly dependent on the category of waste to be stored. Table 2-1 illustrated the classes of landfills in South Africa.

Table 2-1: Landfill classes (Department of Environmental Affairs - South Africa, 2013).

Listed wastes	Landfill Class
i) Asbestos Waste. ▪ Spoilt or unusable hazardous products. ▪ General waste, excluding hazardous domestic waste. ▪ Mixed, hazardous chemical wastes from analytical laboratories and laboratories from academic institutions in containers less than 100 litres.	Class A
i) Domestic waste. ▪ Business waste not containing hazardous waste or hazardous chemicals. ▪ Non-infectious animal carcasses. ▪ Garden waste.	Class B
i) Post-consumer packaging. ▪ Waste tyres.	Class C
i) Building and demolition waste not containing ▪ Hazardous waste or hazardous chemicals. ▪ Excavated earth material not containing hazardous waste or hazardous chemicals.	Class D

2.3.2.1 *Landfill Liner Interface Stability*

The usual objective of an engineered landfill is to maximize the waste disposal capacity in less space while eliminating or minimising the pollution of the surrounding environment. For this reason, the landfill is designed and constructed at very high and steep slopes with geosynthetic installed as containment barriers as a liner (Miuzzi 2012; Stark et al. 2011). Figure 2-14 shows a typical schematic example of an engineered landfill.

The geometry, i.e. in Figure 2-14, however, contributes to slope failures of landfills if the increased elevation heights and steeper side slopes are not compensated to ensure stability (Miuzzi 2012). In fact, these failures are sometimes catastrophic such that properties and lives are lost (Blight 2008; Mitchell et al. 1993).

Figure 2-14: Typical landfill geometry with steep slopes (after Miuzzi 2012).

One of the main concerns when geosynthetics are employed in landfills, especially on the slopes, is their performance when subjected to shear forces. Researchers (i.e. Stark et al., 1998; Bouazza et al., 2002) have shown that the stability of the landfill liner systems is mostly controlled by interface shear resistance mobilised between geosynthetics and the adjacent materials. Usually, an interface slips when the induced shear force above the material is greater than the shear strength below it. This is because when the slope length and angle are increased, the shear forces initiated by gravity are also increased (Miuzzi 2012). Thus, if the shear resistance of the weakest

interface of a liner is not sufficient enough to resist the shear force, the sliding occurs and consequently, the landfill slope fails. Figure 2-15 illustrates the interface sliding/resistance experienced on a steep slope.

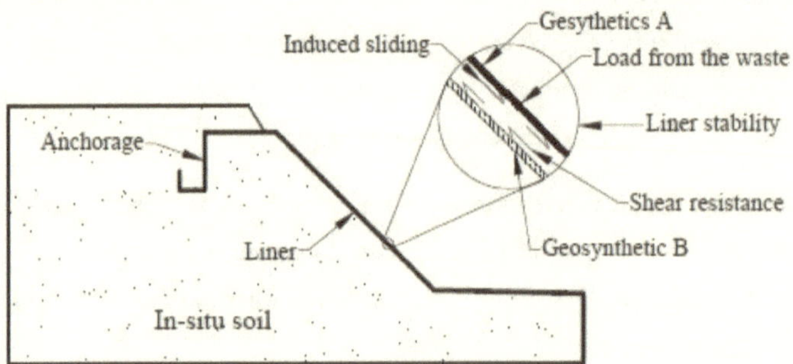

Figure 2-15: Slope stability of the landfill composite liner system (after Miuzzi 2012).

2.3.2.2 Failure Modes of Landfills

The stability of a landfill had sometimes been overlooked in the design stages, as a consequence, some major failures along liner slopes, landfill foundations, and within the waste mass itself have been recorded (Qian et al. 2003). These observed failures can be classified into two main failure modes, namely rotational and translational (Qian et al. 2003). According to Buthelezi (2017), both the rotational and translational failure can involve the waste stream and the subsoil on the foundation. These failure mode types can be further categorized into the following:

- Sliding failure along the leachate collection system
- Rotational failure along sidewall slope and base
- Rotational failure through a waste, liner, and foundation subsoil;
- Rotational failure within the waste mass
- Translational failure by movement along the underlying liner system.

The last category, translational failure along the base with a linear or rotational back-surface failure through the waste, according to Qian et al. (2003), is the most likely failure type to occur in many modern landfills. This group is caused by the existence of liner systems below the waste mass that involve multilayer components consisted of clay soils as well as geosynthetic materials (Qian et al. 2003). Also, it is important to note that the geometry of the landfill, properties of

materials involved, i.e. natural or geosynthetics liners, external loading, leachate levels and drainage conditions can also contribute to the translational failure (Buthelezi 2017).

To ensure the stability of the landfill, a deeper understanding of the characteristics that caused the failure to occur is required. This involves the evaluation of each interface component of a liner on the landfill (Cilliers 2018c). This is because finding the weakest/critical interface that is likely to cause failure, is key in the stability analysis of multilayer lined landfill (Qian & Koerner 2004). Accordingly, various researchers, including Martin & Koerner (1985), Koener and Hwu (1991) and Giroud (1995) proposed numerous analytic methods of assessing the failure in a landfill that involved using a rotational failure method. However, using this type of method to evaluate a translational failure was likely to result in an overestimation of the i.e. slope stability of the landfill due to the corresponding higher factor of safety (a value that estimates how close or far it is from failure as reported by Hammah et al. in 2009) obtained during the analysis (Qian et al. 2003). As a result, Qian et al. (2003) derived a two-part wedge method for translational failure analysis using the conventional limit equilibrium method to evaluate the potential failure systems in a landfill. The method estimates a representative value of the factor of safety (SF) that ensures that the strength of the waste is not exceeded anywhere within the waste mass (Qian et al. 2003).

2.3.2.3 *Translational Failure Analysis of Landfills*

In Figure 2-16, a typical example of a translational failure is illustrated and based on this figure, Qian et al. (2003) derived a two-part wedge method to calculate the factor of safety (FS) for the waste stream against possible translational failure with predetermined sliding failure faces. As seen in Figure 2-17, the waste mass was divided into two distinct parts: an active and passive wedge. An active, according to Qian et al. (2003), is the wedge lying on the back slope that can result from either a lined or previously placed waste and can cause failure. A passive wedge, on the other hand, is the wedge lying on the foundation soil or liner system of the landfill and is likely to resist failure (Qian et al. 2003).

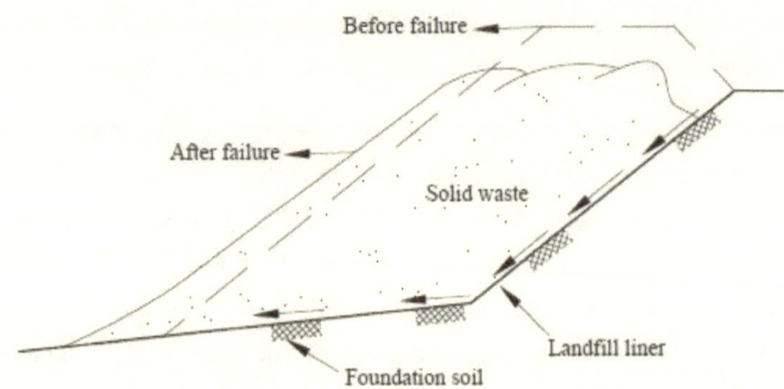

Figure 2-16: Translational waste mass; sliding completely along or within the landfill liner system (after Qian et al. 2003).

In Figure 2-17, Qian et al. (2003), assumed that the resultant of inter wedge force, ω was inclined at an unknown angle to the normal drawn to the interface between active and passive wedges. Also, the line of action for this resultant force act at a distance of H/3 above the base of the interface (Qian et al. 2003). This meant that the FS at the interface between active and passive wedges should not be less than 1.0 (Qian et al. 2003).

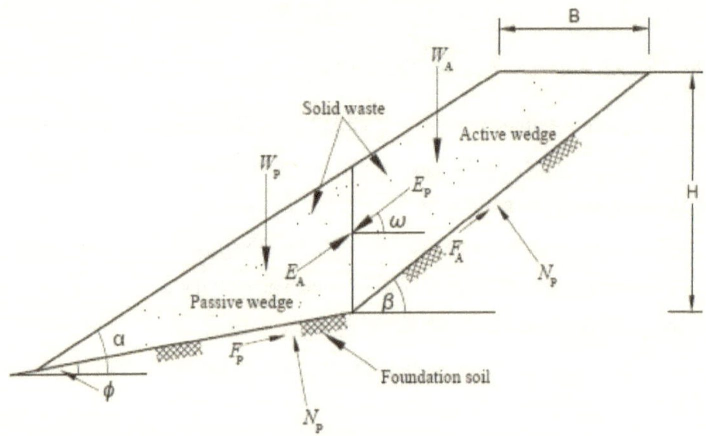

Figure 2-17: Translational waste mass; forces acting on two adjacent wedges of a waste mass (after Qian et al. 2003).

Considering the equilibrium of the whole waste mass in Figure 2-17, Qian et al. (2003) formed equation 2-1 to calculate the minimum FS_{min}. In this study, $FS_{min} = FS$ and it was assumed to be the same at all points of the failure surface (Qian et al. 2003). The additional assumption made in the derivation of the equation 2-1 can be found in the publication entitled "Translational Failure Analysis of Landfills" by Qian et al. (2003).

$$FS_{min} = \frac{-b \pm \sqrt{b^2 - 4 \times a \times c}}{2 \times a}$$

Equation 2-1

Where a, b and c can be obtained as:

$$a = W_A.\sin\beta.\cos\theta + W_p.\cos\beta.\sin\theta$$

Equation 2-2

$$b = \left(W_A.\tan\delta_p + W_p\tan\beta\right)\sin\beta.\sin\theta - \left(W_A.\tan\delta_a + W_p.\tan\delta_p\right)\cos\beta.\cos\theta - C_A.\cos\theta - C_p.\cos\beta$$

Equation 2-3

$$c = -\left[\begin{array}{l}\left(W_A\cos\beta.\sin\theta + W_p.\sin\beta.\cos\theta\right)\tan\delta_a.\tan\delta_p \\ + C_A.\sin\theta.\tan\delta_p + C_p.\sin\beta.\tan\delta_a\end{array}\right]$$

Equation 2-4

The values of W_A, W_P, C_A and C_p can be calculated using the following equations:

When $B < \dfrac{H}{\tan\beta}$,

$$C_A = c_a.\frac{H}{\sin\beta}$$

Equation 2-5

$$W_A - 0.5.\gamma_{sw}.\frac{H^2}{\tan\beta} - 0.5.\gamma_{sw}.\left(\frac{H}{\tan\beta} - B\right)^2.\tan\alpha$$

Equation 2-6

$$C_P = c_p\left[\left(H - \frac{H.\tan\alpha}{\tan\beta} + B.\tan\alpha\right) \div \left(\cos\theta.\tan\alpha - \sin\alpha\right)\right]$$

Equation 2-7

$$W_p = 0.5.\gamma_{sw}.\left[\left(\frac{H}{\tan\alpha} - \frac{H}{\tan\beta} + B\right)^2.\frac{\tan\alpha.\tan\theta}{\tan\alpha - \tan\theta} + \left(\frac{H}{\tan\alpha} - \frac{H}{\tan\beta} + B\right)^2.\tan\alpha\right]$$ Equation 2-8

When $B \geq \dfrac{H}{\tan\beta}$,

$$W_A = 0.5.\gamma_{sw}.\frac{H^2}{\tan \beta} \qquad \text{Equation 2-9}$$

$$W_p = 0.5.\gamma_{sw}.\left[\left(\frac{H}{\tan \alpha} - \frac{H}{\tan \beta} + B\right)^2.\frac{\tan \alpha.\tan \theta}{\tan \alpha - \tan \theta} + H\left(B - \frac{H}{\tan \beta} + \frac{0.5H}{\tan \alpha}\right)\tan \alpha\right] \quad \text{Equation 2-10}$$

It is significant, however, to note that obtaining an accurate factor of safety in the translational failure analysis, the correctly determined field and/or laboratory input parameters (i.e. shear strength properties) are absolutely critical (Koerner & Soong 2005). This is because the accuracy of the final analysis is only as good as the correctness of the field and/or laboratory obtained input values (Koerner & Soong 2005). In this study, the determination of the shear strength characteristics in the laboratory was considered.

In laboratory test, a direct shear box, ring shear device, tilt table or cylindrical shear device can be used to determine the geosynthetic shear strength properties as per site anticipated condition (Bouazza et al. 2002). The direct shear box, however, is the most reliable apparatus utilized mainly in accordance with ASTM D5321 and ASTM D6243 standards (ASTM D5321 2013; ASTM D6243 2018; Bouazza et al. 2002). The ASTM D5321 provides the methods for investigating the interface shear strength of all geosynthetics except for GCL (ASTM D5321 2017). The determination of all GCL interface shear behaviour is prescribed in ASTM D6243 which is similar to ASTM D5321 but differs in specimen confinement, specimen preparation procedures and shear displacement rates (McCartney & Swan 2002).

2.3.3 Determining the Interface parameters

To verify the stability of the landfill liner, the interface shear parameters between i.e. two geosynthetics are analysed. These parameters are mainly determined from the laboratory test performed in a shear apparatus, with the interface of the geosynthetic specimens lying in the plane of shear. The test device, whose primary purpose is to measure the maximum and ultimate interface shear strength between the tested geosynthetics for various normal stresses, is a modification of a conventional standard direct shear machine used to determine the angle of internal friction of soils (Kalumba 1998). A detailed description of the test method, analysis and results presentation for the determination of the interface friction angle was given in the following section.

2.3.3.1 *Direct Shear Apparatus*

A direct shear device is typically divided into two shearing blocks, the lower/bottom moving part, and the upper/top static part as shown in Figure 2-18. A constant normal force, N is applied to the top box as illustrated in Figure 2-18 to produce a vertical normal stress, $\sigma = N/A$, where A is the cross-sectional area of the direct shear apparatus. A steadily increasing shear force, F which causes an increasing shear/horizontal displacement is applied mainly to the bottom part of the shear box at a user-defined rate (i.e. 0.1 mm/min, 1.0 mm/min or 1.5 mm/min) such that the lower section moves laterally relative to the upper static part initiating the shearing process. Once the shearing process is initiated, the relative horizontal displacement of the bottom shearing block to the top shear box and the applied normal force is monitored by the test machine through a connected desktop computer. As a result, a respective shear stress versus horizontal displacement relationship graph is plotted, as illustrated in Figure 2-19.

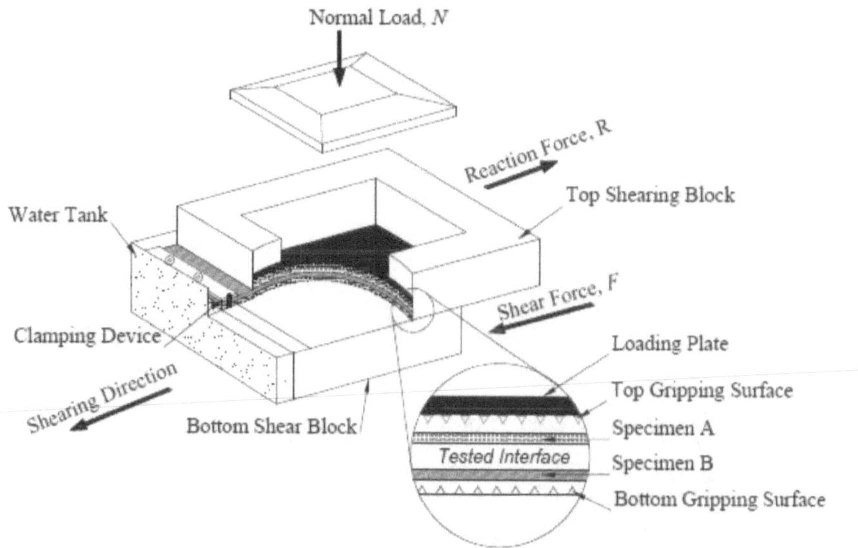

Figure 2-18: Large Direct Shear Apparatus.

Generally, the horizontal displacement of the tested geosynthetic takes place gradually with the increase in the application of the normal stress. The maximum shear strength, also referred to as peak strength is then achieved after a certain amount of displacement as indicated at point A in Figure 2-19. This is then followed by a decrease in shear strength as shown by region BC in

Figure 2-19 to reach the residue shear strength. At this stage, the tested material is going through a plastic deformation state, a process termed as strain softening (Buthelezi 2017). Once the residual strength is reached, the tested specimen is considered to have 'failed' and the experiment is terminated by the user or ends automatically (Jogi 2005).

It is, however, important to note that when the 300 mm x 300 mm direct shear device is used, the residue shear strength condition may not be defined in some cases due to its limited shear stress-displacement (Rouncivell 2005). Therefore, a reduced shear strength condition reached after the strain softening is referred to as the large displacement (LD) strength (Rouncivell 2005).

At the completion of the test, the failure mode of the tested specimen is carefully inspected to ensure that failure happened at the intended interface and the stress calculations took the actual sheared area into consideration (Rouncivell 2005). This test procedure, mentioned, is then repeated for at least three times by applying different normal stresses to obtain a set of three points in the shear stress versus normal stress graph, which identify the Mohr-Coulomb failure envelope (ASTM D5321 2017; ASTM D6243 2018).

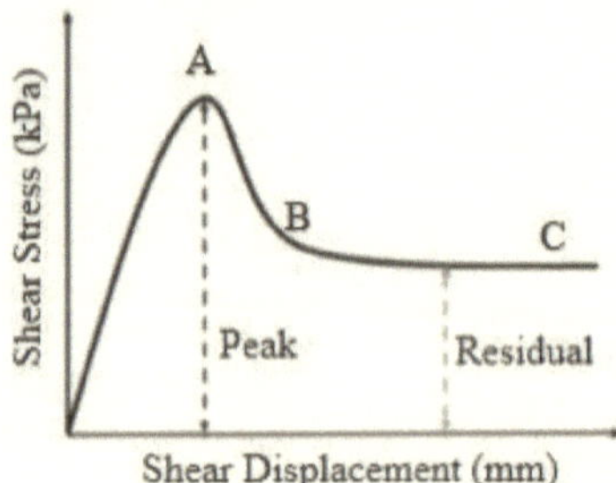

Figure 2-19: Shear stress versus horizontal displacement relationship

2.3.3.2 *Mohr-Coulomb Failure Envelope*

In 1900, Mohr presented a theory stipulating that the failure of material was influenced by a critical combination of normal stress and shearing stresses. This criterion is now mostly used in many geotechnical analysis methods and programs to define the shear strength of materials including geosynthetics. The Mohr-Coulomb criterion, which is represented by equation 2-11 describes a linear relationship between the shear and normal stresses at maximum and residual strength.

$$\tau = a + \sigma_n \tan \delta \qquad\qquad \text{Equation 2-11}$$

Where τ = Shear strength

a = Adhesion corresponding to the inclination of the vertical axis

σ_n = Applied normal stress

δ = Internal/interface friction angle corresponding to the inclination of the horizontal axis

In Figure 2-20, the shear stress versus normal stress graph, is plotted using the normal stress values at peak strength with their corresponding shear strength read-off from the shear stress vs horizontal displacement curve like the one shown in Figure 2-19. This graph basically approximates a "best-fit" straight line through a defined section of the shear stress versus normal stress failure envelope described by the Mohr-Coulomb equation, as shown in Figure 2-20. The inclination of the line to the normal stress axis is interpreted as friction angle and its intercept with the shear stress axis interpreted as the apparent adhesion (Jogi 2005; ASTM D5321 2017). Thus, if the shear stress versus horizontal displacement curve does not represent the actual performance of the tested geosynthetics, the shear stress values used to plot the shear stress versus normal stress graph will be incorrect. Accordingly, the design parameters will not be accurately representing the field performance of the material.

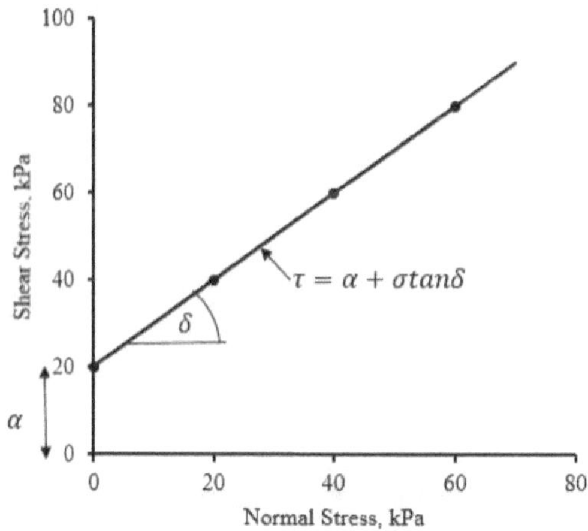

Figure 2-20: Shear Stress vs Normal Stress.

The Mohr-Coulomb theory, according to Juvinal & Marshek (1991), generally applies to materials with the compressive strength that far exceeds the tensile strength, and this might not be the case with geosynthetics especially if loaded at high stresses (Kalumba 2018). Jogi, (2005) in his study highlighted the importance of analyzing the nonlinearity nature of the shear stress versus normal stress graphs of geosynthetics, especially if tested over a wide range of normal stresses.

In Figure 2-21, a nonlinear response of the shear stress versus normal stress test results is shown. The graph illustrated how 'approximate' linear relationships are often used to describe the failure envelopes, particularly for geosynthetics. The straight line (1) provides a reasonable fit for high normal stress values, but significantly overestimates the low-stress response (see Fig.2-21a). Conversely, using the failure envelope represented by line 2 in Figure 2-21a, it will provide a satisfactory approximation at low normal stress values but severely overestimates the high-stress response of the material. In such cases, the ASTM D5321/6342 recommends that a best-fit line (3) is to be used by the designer to describe the failure envelope of the material. However, selecting line (3) may lead to either an overestimate or underestimate of the actual performance of the material, depending on the range of the stresses to be encountered (Giroud et al. 1993). Therefore, to accurately assess the actual laboratory response of such material (in Fig. 2-21), a bilinear envelope using line (1) and (2) can be used as shown in Figure 2-21b.

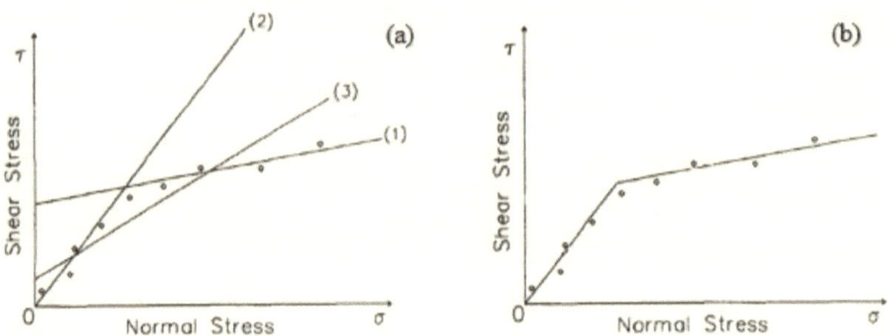

Figure 2-21: Various failure envelope fitting methods (after Giroud et al. 1993)

2.3.3.3 *Effect of Normal Stress*

The ASTM D5321/D6243 standards state that a best-fit straight line governed by the Mohr-Coulomb failure envelope can be developed if at least three test points are conducted at different normal stresses in geosynthetics interface shear tests. However, researchers including McCartney and Swan in 2002 have found that the application of normal stress during geosynthetic interface shear strength testing can affect the shear strength results obtained. McCartney and Swan in 2002, in their study reported that the magnitude of the shear strength of the GCL increases with increasing normal stress as represented by the Mohr-Coulomb failure criterion. According to this criterion, the friction angle defines a linear increase shear strength with increasing effective normal stress (McCartney & Swan 2002). The downside with this is that the Mohr-Coulomb linear failure envelope may not sufficiently represent the failure envelope experienced by geosynthetics as their shear strength may be non-linear over a wide range of normal stresses (Jogi 2005).

2.3.3.4 *Shear Displacement Rate*

Shear displacement rate is an important aspect in geosynthetic interface shear strength testing. Previous studies (i.e. Fox & Stark 2004; Eid et al. 1999; Zornberg et al. 2005; Stark et al. 2015; Zelic et al. 1998; McCartney & Swan 2002) have shown that the choice of the shear displacement rate affects the measured interface shear strength of geosynthetics (i.e. GCLs). For example, Zelic et al. (1998) reported that the displacement rate values of 1,219 mm/min and 0,00146 mm/min in GCL shear strength testing has a reduction of approximately 35% of the peak values and about 52%ofr the residual values, respectively. Also, the shear rate value can affect the time and cost perspective, especially with regard to commercial testing, (McCartney & Swan 2002; Rouncivell 2005). On contrary, Ross et al. (2010) reported that the interface shear strengths of geosynthetics (i.e. geomembrane/GCL) are generally not affected by the displacement rate at low to moderate normal stress levels, (normal stress less than 629 kPa).

Generally, the shear rate in the field is anticipated to take place at a slow rate (Zornberg et al. 2005). If a faster rate is applied during lab tests, the quality assurance of the test performed may be affected as the faster rate may promote tension or slippage of tested geosynthetic specimens. Previous researchers, including Zelic et al. (1998); McCartney & Swan (2002) and Stark et al. (2015) have used and recommended different shearing rate i.e. 0.01, 0.1 and 1.0mm/min.

However, a shear rate of 0.1 mm/min and/or 1.0mm/min is mainly utilized in geosynthetics interface testing (McCartney & Swan 2002; Stark et al. 2015; Buthelezi 2017; Fox & Stark 2004).

2.3.3.5 *Hydration*

Hydration considerations are very important when conducting interface shear strength testing of geosynthetics, especially for GCL. The GCL is a unique geosynthetic material that contains bentonite material which is a natural mineral with swelling characteristics. If the GCL is to be applied in the landfill, its stability consideration must be checked. This is because when the bentonite is hydrated (by rain or moisture from waste), the shear strength of the GCL decreases and can affect the stability of the slope (Fox & Stark 2004). Therefore, if hydration is anticipated in the field, which is usually the case, laboratory tests should be performed under hydrated conditions to represent the least favourable environments (Fox & Stark 2004).

The ASTM D6243 standards recommends that the hydration of the GCL test specimens should be carried out under the normal stress anticipated in the field at the time of hydration. For this reason, researchers, including Eid and Stark (1996; 1997) hydrated Bentofix GCL specimens under a normal stress of 17kPa to represent the field conditions anticipated before waste placement. The hydration phase lasted for three to four days to achieve full hydration of the GCL (Eid & Stark 1996). Although full hydration represents the most critical state of GCL in the landfill liner, dry or partial hydration is the most common state of GCL at work (Lin et al. 2014). Most production testing laboratories hydrate GCL for 24 to 48 hours because a little decrease in shear strength is expected if shearing occurs at the same normal stress (Fox & Stark 2015; McCartney et al. 2009). Moreover, the ASTM D6243 does not provide the specific time for hydration.

The geotextile/geomembrane interface test, on the other hand, can be hydrated for an hour as no soil undergoes consolidation. The one hour is to allow sufficient time for the geotextile and geomembrane to engage prior to shearing (Stark et al. 2015).

2.3.3.6 *Consolidation Time*

Laboratory tests have shown that specimen consolidation during the interface shear test can negatively affect the determination of the interface shear strength of geosynthetics (i.e. GCLs)

(Fox & Stark 2015). Thus, to obtain the actual field performance of the tested GCL, a sufficient consolidation time frame is required before shearing.

Jogi (2005), recommended that the interface shear strength test is to be conducted immediately after the application of the normal stresses (unconsolidated loading) to approximately model the relatively rapid placement of waste during landfilling operations. Conversely, allowing an adequate consolidation time under the imposed normal load before conducting an interface shear strength test can simulate the long-term response of the liner (Jogi 2005).

It is, however, important to note that a full consolidation time of the test GCL specimens is not certain during interface shear testing (Jogi 2005). Thus, researchers have used different consolidation time frame including 24 hours, 1 hour or less (Fox & Stark 2004). The problem with such short consolidation time is that inaccurate shear strengths could be produced due to bentonite extrusion in the GCL and the presence of positive pore pressures in the specimens at the start of shearing (Fox & Stark 2004).

2.3.3.7 *Specimen Gripping Systems*

In geosynthetic direct interface shear testing, the test specimens are secured to the lower and upper shearing block using a specimen gripping system. This system is a combination of the gripping surface and a clamping device.

The gripping surface provides adequate resistance between the test samples and the shearing blocks such that the specimens do not slip during shearing. This is to simulate the anticipated field interface resistance between two geosynthetics or geosynthetic and the adjust material (Kalumba 1998). Generally, the grip may contain short, sharp pins or teeth that "bite" into the geosynthetics to mobilize sufficient resistance against slippage (Fox et al. 2004). However, many gripping surfaces used for geosynthetics testing are not adequately aggressive to shear strong materials such the reinforced GCL without assistance, hence, clamping devices are used to facilitate the shearing in most testing laboratories (Fox et al. 2004). A clamping device which can represent the anticipated field edge anchorage, offer extra strength to hold the tested samples in position and enforce failure of the specimen to occur at the desired interface (Fox et al. 2004).

In overall, if the specimen gripping system is not used, the tested geosynthetics slips on the shearing blocks during shearing, generating tensile forces within the specimen, hence, causing progressive failure of the tested specimen. This progressive failure reduces the measured peak shear strength and increases the large displacement shear strength (Fox et al. 2004) as part of the applied normal load is not transferred to the tested samples. This causes the determination of shear strength parameters that do not represent the actual field performance of the tested geosynthetics, and consequently, the design is negatively affected. Therefore, it is of importance to provide a specimen gripping system that secures the test specimens in position, prevent slippage between the specimen and the shearing block and does not interfere with the measuring of shear strength parameters of the tested specimen during shearing.

Currently, the ASTM standards state that the textured surface and flat jaw-like clamping device are normally sufficient but work is still in progress to define the best type and textured surfaces that can be used to minimise or eliminate errors involved with the specimen gripping (ASTM D5321 2017; ASTM D6243 2018).

2.3.3.7.1 *Gripping Systems Used*

Over the years, researchers have come up with different gripping systems that can be used to determine the interface shear strength of geosynthetics. These systems are mainly a combination of the metal clamping device with the textured steel, adhesive bonding glue, nail plate or sandpaper. These gripping systems, however, are sometimes reported to be not adequate to sufficiently secure the tested specimens to the shearing device. This has led to a large variability in test results and difficulties in their interpretation. The following commonly used gripping systems were discussed:

Textured Steel

Researchers, including Merrill and O'Brien (1997); Swan et al. (1997); Trauger et al. (1997); Zornberg et al. (2005) and (Fox et al. 1997) have used a textured steel gripping surface in geosynthetic shear testing. During the test, the specimens were secured both to the upper and lower shearing block using the steel grips. The grip surfaces were made up of parallel woodworking rasps that were uniformly distributed such that they would create an interlocking bond with the specimens once the loading stress was applied, thus, providing adequate surface resistance against slippage (Fox et al., 1997).

Zornberg et al. (2005) also, secured the GCL to the shearing blocks using a similar gripping system, though a porous rigid substrate surface was utilized in this case as shown in Figure 2-22. According to the authors, this type of gripping system allowed minimal to zero slippage to take place during shearing such that clamping device was not used. As a result, a relatively uniform transfer of shear load to the tested specimen was achieved, thus, shear failure occurred at the intended interface (Swan et al. 1997; Trauger et al. 1997; Zornberg et al. 2005; Fox et al. 1997).

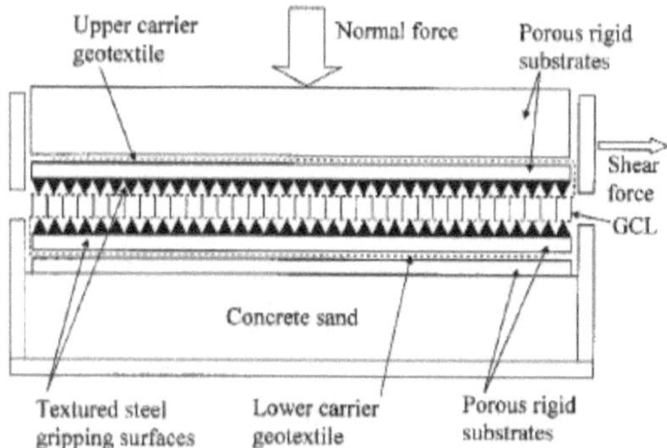

Figure 2-21: Specimen gripping detail in direct shear (Zornberg et al. 2005).

Adhesive Bonding

The use of adhesive bonding is also another geosynthetic gripping method that has been reported, (David & Robert 1991; Eid et al. 1999; Fox & Ross 2011; Triplett & Fox 2001).

Polyvinyl chloride (PVC) and the water-resistant epoxy solvent have been used as gripping surfaces to bond the geosynthetics to the bottom direct shearing blocks (David & Robert 1991; Fox & Ross 2011). The adhesive bonding was used together with the clamping device to improve confinement of the specimen to the bottom platen (David & Robert 1991).

In a similar study conducted by Eid et al. (1999) and Triplett and Fox (2001) to determine the interfacial and internal shear strength parameter for the geosynthetics, glue was used as a gripping surface. Eid et al. (1999), used the adhesive to bond the test geosynthetics on the modified Bromhead torsional ring shear apparatus before initiating the shear test. Triplett and Fox (2001), on the other hand, used water-resistant spray-on adhesive and a two-part epoxy to secure the tested geosynthetics on the pullout shear machine in order to mobilize the shear resistance between the test sample and the shearing block.

The adhesive employed during the studies, according to David and Robert (1991), Eid et al. (1999), Fox and Ross (2011), Triplett and Fox (2001), it produced uniform shear displacement for each tested specimen indicating that slippage was minimal or prevented. However, Triplett and Fox (2001) reported that this method was limited to lower normal stresses of less than 280 kPa.

Despite the great findings by other researchers (i.e. David & Robert 1991; Eid et al. 1999; Fox & Ross 2011), Fox et al. (2004) stated that glue is not to be used in shear strength testing of geosynthetics especially when working with GCL as there are possible interference with the failure systems if careful steps are not taken.

Figure 2-23 shows the specimen configuration used by David and Robert (1991).

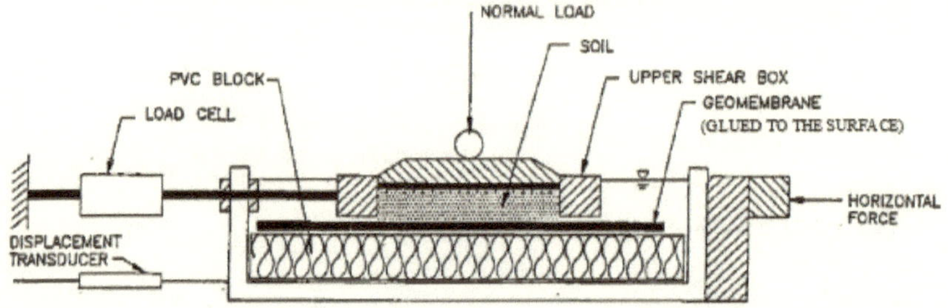

Figure 2-22: Specimen gripping detail in direct shear (after David and Robert, 1991).

Nail Plates

A nail plate is also another method of avoiding slippage during shear tests on a direct shear box. Zanzinger and Alexiew (2000) investigated the shear strength of GCL using a nail plate moulded in epoxy resin as a gripping surface. The nail plate was used together with a metal end clamp to hold the tested specimen as shown in Figure 2-24. The plate surface, according to Zanzinger & Alexiew (2000), was made up of 2 mm high nails, which were evenly distributed at one nail per square centimetre (cm). During the test preparations, the nail plates were placed on both the upper and lower shearing blocks to hold the GCL such that the two nail plates did not touch each other at any time during shearing (Zanzinger & Alexiew 2000). This method of specimen gripping provided adequate resistance against slippage, consequently, transferred the applied stresses across the tested specimen (Zanzinger & Saathoff 2010). This led to Zanzinger and Saathoff to adopt the method and improve it by using stainless-steel nail plate with a better drainage (Zanzinger & Saathoff 2010).

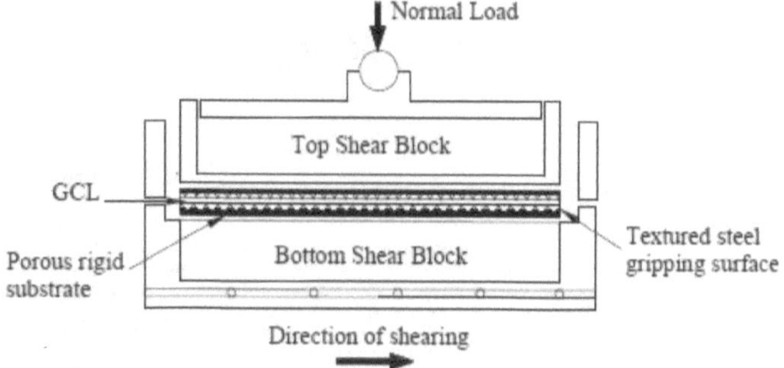

Figure 2-23: Specimen gripping detail in direct shear (after Zanzinger and Alexiew 2000)

Sandpaper

Substantial studies have been conducted on the shear strength behaviour of geosynthetic interfaces using several gripping systems (i.e. textured steel and glue), however, very few have been published on sandpaper been a gripping surface. Fox et al. (1997) conducted a study on the design and evaluation of a large direct shear machine for GCLs, were two different surface size of sandpapers were used. The two types of the sandpapers used were the medium coarse and coarse wet/dry industrial sandpapers. The author used the sandpapers to create resistance against slippage between the GCL and the shearing block. The clamping device was also employed at one end to provide extra support (Fox et al. 1997). According to Fox et al. (1997), both the gripping systems did not provide adequate resistance to prevent slippage during shearing. In Figure 2-25, the test configuration of the experiment is shown.

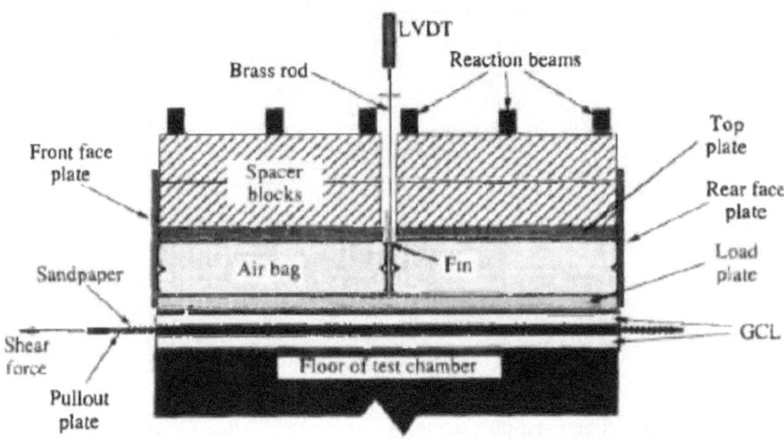

Figure 2-24: Specimen gripping detail in pullout (Fox et al., 1997).

Summary of gripping system

In Table 2-2, a summary of the reviewed gripping system for geosynthetics interface testing using different types of direct shear apparatuses was presented.

Table 2-2: Summary of the gripping system

Author	Test Setup	Gripping System used	
Merrill and O'Brien (1997);	Direct shear box		
Swan et al. (1997);	Direct shear box		
Trauger et al. (1997);	Direct shear box	-	Textured Steel
Zornberg et al. (2005)	Direct shear box (305 mm x 305 mm)		
Triplett and Fox (2001)	Pullout Shear Machine		
Rouncivell, (2005)	Direct shear box (60 mm x 60 mm)	Metal bar	
David and Robert, (1991)	Direct shear box (60 mm x 60 mm)		
Eid et al., (1999)	Ring shear machine	-	Adhesive Bonding
Triplett and Fox (2001)	Pullout Shear Machine		
Fox and Ross (2011)	Direct shear box		

Continuation of Table 2 2: Summary of the gripping system

Author	Test Setup	Gripping System used	
Zanzinger and Alexiew (2000)	Direct shear box (300 mm x 300 mm)	-	Nail Plates
Zanzinger and Saathoff (2010)	Direct shear box (300 mm x 300 mm)		
Buthelezi (2017)	Direct shear box (305 mm x 305 mm)	Metal bar	
Fox et al. (1997)	Pullout Shear Machine	Metal bar	Sandpaper
Kalumba (1998)	Direct shear box (100 mm x 100 mm)		

2.3.3.7.2 *Effects of Gripping Systems*

Fox et al. (1997) conducted a study where the effects of three different gripping systems (truss plate, coarse and medium sandpaper) were studied. The author sheared a hydrated woven (W)/non-woven needle-punched (NW NP) GCL at a constant normal shearing stress of 37.8kPa. Figure 2-25 shows the comparison of the three results obtained during the shear tests. For the first test, the truss plate without end clamping was used and Figure 2-25, shows a measured uniform shear stress versus horizontal displacement curve of the tested GCL specimen. A second test was performed to investigate coarse wet/dry sandpaper gripping surface with an end clamp system clamping the geotextile. The third gripping system used in their study was a medium coarse wet/dry, industrial strength sandpaper clamped on the edge of the shear box.

Comparing the results obtained using the three gripping systems, the truss plates without end clamping produced a uniform shear failure with a higher peak of 75.9 kPa at a smaller displacement and marginally lower residual strength of 5.7 kPa than the coarse and medium sandpaper. The coarse sandpaper produced a well-defined shear failure peak of 65.7 kPa at a peak displacement larger than that of the truss plate and a residual strength of 6.9 kPa (Fox et al. 1997). The medium coarse sandpaper, however, did not mobilize sufficient friction to prevent tensile failure, hence, slippage occurred rendering the test invalid (Fox & Stark 2004).

Based on the mobilised peak strength between the truss plate and coarse sandpaper, a percentage difference of approximately 13% was evident. This showed that the type of gripping systems used on shear strength at geosynthetic interface has an effect on the accuracy of the results obtained.

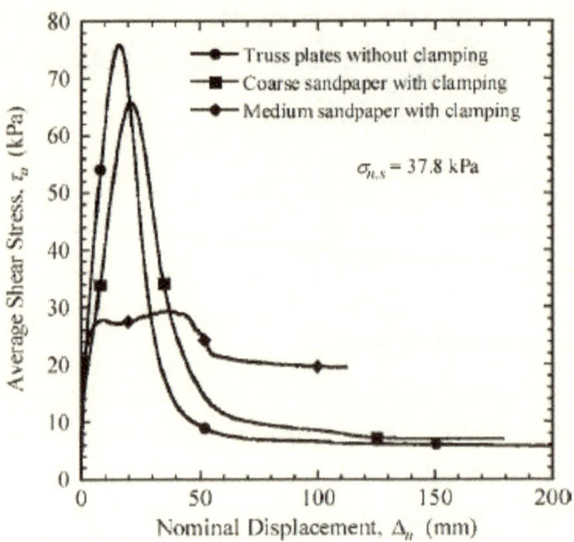

Figure 2-25: Shear stress versus horizontal displacement curve, (Fox et al. 1997, Figure 8).

In a separate study, Fox et al. (2004) presented shear stress versus horizontal displacement relationship curves for shear strength of hydrated needle-punched (NP) GCLs obtained using textured gripping systems and compared them to the results determined by Fox et al. (1998) when a modified metal connector plates without end clamps was utilized.

Generally, a shear stress versus horizontal displacement curve representing the true performance of the material shear behaviour has a similar shape to one in Figure 2-26a (Fox 2010). The curve exhibits smooth transitions from the start of loading to peak shear strength, strain softening and then to large displacement/residual shear strength, (Fox 2010). The relationships curves obtained from replicate tests should show good similarity as normal stress is increased or decreased.

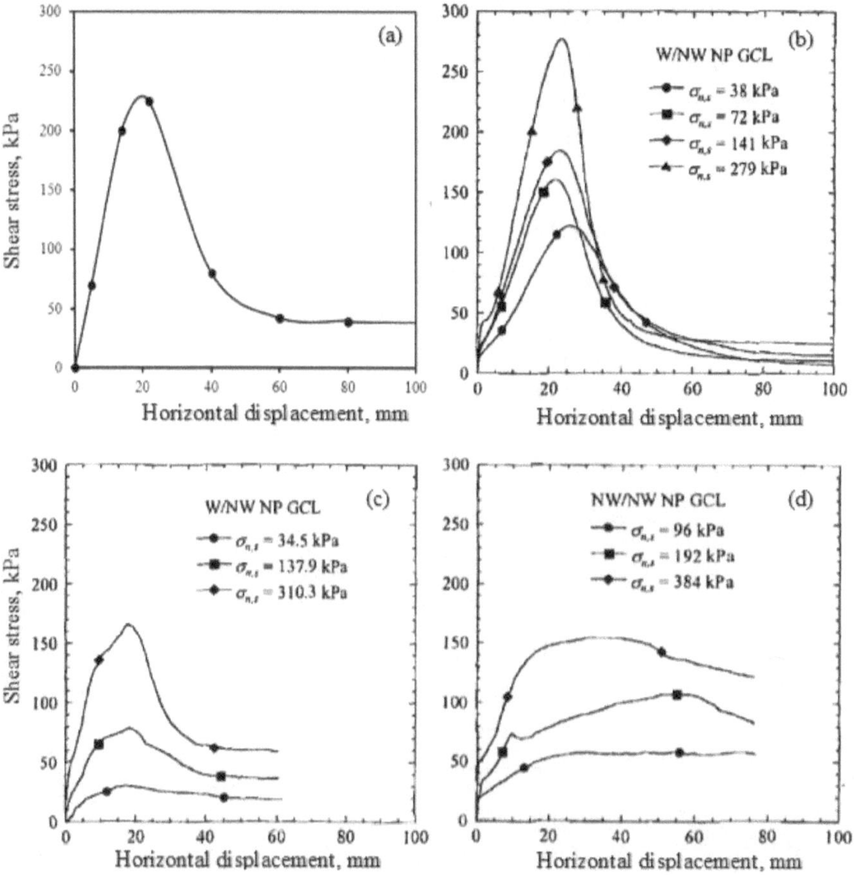

Figure 2-26: Shear stress versus horizontal displacement curves - NP GCL; (a) Typical shear stress versus horizontal displacement curve, (b) Modified metal plate with no clamps, (c) Modified metal plate with clamps, and (d) Modified metal plate with clamps but the but the problem had occurred, (after Fox et al., 2004).

The shear stress versus horizontal displacement curves in Figure 2-26b were the results obtained using the metal plate gripping surface without end clamps. The relationship curves displaced an increase in the shear strength with increasing applied normal stress to reach sharp, narrow peaks at lower displacements (see Fig. 2-26b). Beyond the peaks, the plots decreased in shear stress to approach the residual shear strength of the tested geosynthetics. According to Fox et al. (2004),

the graph displayed in Figure 2-26b are most likely the accurate representation of actual material shear behaviour. This was confirmed by comparing high-quality shear stress versus horizontal displacement relationships provided by Eid et al. (1999) for torsional ring shear and by Triplett and Fox (2001) and (Fox et al. 1998) for direct shear, which displayed a strong similarity in the shear mobilizations.

Also, Figure 2-26c presented the shear stress responses investigated on a textured steel gripping surface with end clamps. These relationships indicated slightly wider peaks at lower stresses of 34.5 kPa and 137.9 kPa. It can be further noticed that at high pressure of 310.3 kPa, a small stress peak achieved. Figure 2-26d, on the other hand, displaced double peaks, poor similarity, an absence of post-peak strength reduction ($i.e.\,\sigma_{n,s} = 96\,kPa$) and undulations that were non-physical indicating that a problem occurred during shear testing. According to Fox et al. (2004), the inaccurate relationships presented in Figure 2-26d were most likely due to poor gripping systems that had caused slippage during shearing.

The shear stress versus horizontal displacement relationship for eight direct shear tests conducted on the hydrated textured geomembrane (GMBX)/GCL specimens at four different normal stresses (30.9, 41.2, 51.6, and 61.9 kPa), is presented in Figure 2-27. The tests were performed by Fox and Kim (2008) to compare the shear strength results to a simulated numerical model.

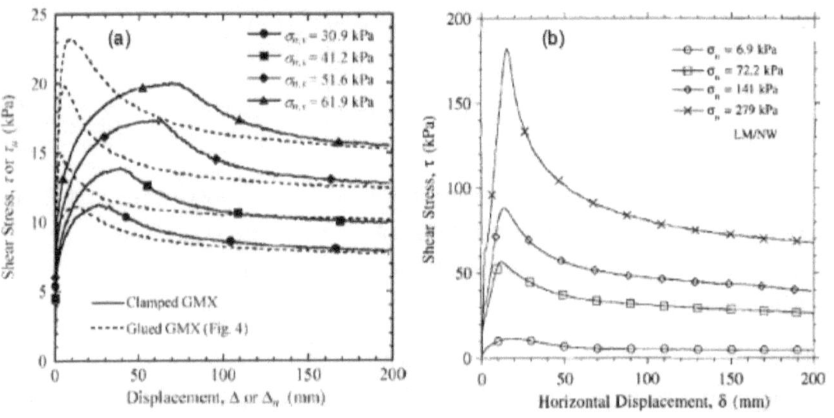

Figure 2-27: Effects of gripping systems on GMBX/GCL interface; (a) Clamped vs glued specimen, (Fox & Kim 2008) and (b) Laminated texture HDPE/Non-woven GCL, (Triplett & Fox 2001).

Fox and Kim (2008) conducted the first four shear tests using glue to bond the tested GMBX specimen to the upper shearing block of the direct shear device. The last four tests, however, the GMBX specimens were clamped at the end and allowed to slip on the shearing block (smooth gripping surface).

Comparing the results obtained in Figure 2-27a to the high-quality shear stress versus horizontal displacement curves presented by Fox et al. (1998); Eid et al. (1999); Triplett and Fox (2001) (see Fig. 2-27b), it can be observed that the relationships investigated when GMBX was glued provided a close range of the true behaviour of the GMBX/GCL interface.

Fox and Kim (2008) also compared the shear stress versus horizontal displacement relationships for glued GMBX/GCL interface tests to the simulated model as shown in Figure 2-28. These relationships, as seen in Figure 2-28a, displayed similarities of the mobilized shear strengths at over certain ranges.

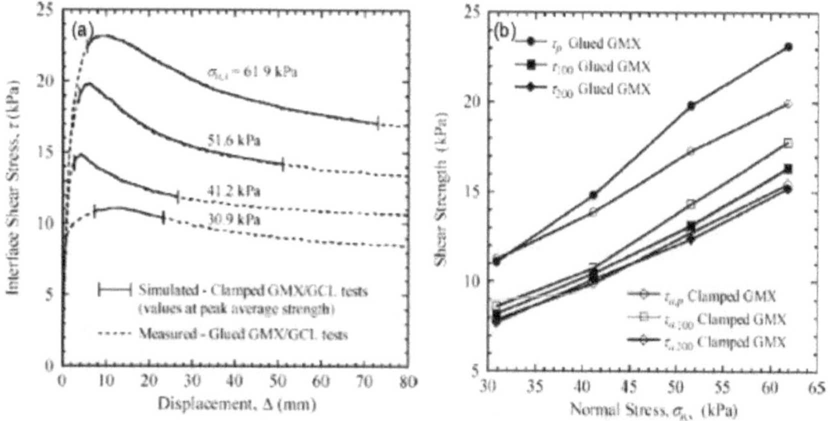

Figure 2-28: Effects of the gripping systems in GMB/GCL interface testing; (a) Glued and Simulated specimen, and (b) Failure envelopes- glued and clamped specimen, (Fox & Kim 2008).

In Figure 2-28b, comparisons of failure envelopes for the peak and residual shear strength of the glued GMBX specimen with the clamped (smooth gripping surface) GMBX specimen is presented.

Observations in Figure 2-28b, indicated that at the lower normal stress the gripping system had no effect on the mobilized peak strength as both glued and clamped GMBX had similar failure envelopes. However, as normal stresses were increased, the glued GMBX test specimen strength envelopes diverged displacing higher peak strength values, as seen in Figure 2-28. According to Fox & Kim (2008), the largest percentage error experienced at the highest normal stress in shear mobilization for the clamped GMBX tests was 14%.

2.3.4   Conducting the Interface Shear Test

Reviewing the relevant literature (i.e. Eid and Stark, 1996; Russell et al., 1998; Fox and Kim, 2008), one can recognize that the interface shear behaviour of geosynthetics had been studied mainly using the single interface tests. The disadvantage with this approach is that it can lead to an overestimation of the shear strength of some geosynthetic interfaces (Stark et al. 2011). It can also consume a substantial amount of time and cost to fully understand the shear strength characteristics of the whole liner system, as a composite liner structure has many interfaces that require testing (Stark et al. 2011).

According to Stark et al. (2011), the failure surface in the single interface test is forced to occur along a specific interface of a geosynthetics which is not the true representation of the field condition. This is because the shear movement in the field may occur on more than one interface plan in a composite liner system, (Stark et al. 1998). These all limitation can be possibly be captured in a multi-interface test, (Stark & Choi 2004).

A properly planned and performed multi-interface tests can be cost and time effective. It can lead to a better interpretation of the interface shear behaviour and potential weaknesses in a composite liner system, (Fox & Stark 2004). In a multi-interface test, interfaces are tested simultaneously and allow failure to occur along the weakest interface as anticipated in the field. The peak and residual design strength are determined directly from multi-interface tests instead of developing combination peak and residual failure envelopes from the results of several single interface tests, (Stark & Choi 2004)

Disadvantages of Multi-Geosynthetic Interface Tests

Multi-geosynthetic interface tests can provide a better simulation of field conditions, however, the tests are limited in that only strength behaviours for the critical surface are provided and not for the other interfaces, (Fox & Stark 2004). Hence, there is no information obtained on how close the other interfaces were to failure, however, has no consequence if the interfaces

determined are a true representation of the field conditions (Fox & Stark 2004). But if there is uncertainty in the test data, which is always the case, then knowing that the interface with a significantly lower residual strength (i.e. hydrated GCL) almost failed could be of importance, (Fox & Stark 2004). Moreover, multi-geosynthetic interface tests, respectively, require that the engineer and testing laboratory have more experience to perform and interpret the tests to avoid errors, (Fox & Stark 2004; Stark et al. 2011). Therefore, it is beneficial to compare the multi-interface test results with at least one test on a single interface, such as the critical interface for a given normal stress because the most experience is with single interface tests and it has fewer variables, (Stark et al. 2011).

In this study, both single and multi-interface tests were conducted to assess the viability of a multi-interface test in composite liner system interfaces.

2.4   Conclusions of the Literature Review

The literature review presented above focused on the topics that are relevant to this research. The important conclusions from the reviewed studies are summarized below:

1. The large direct shear box is the most popular device used for interface shear testing for geosynthetic materials. This is because it can be used for any type of geosynthetic products, a large range of normal stress is possible, large specimens can be tested, the post-peak response can be obtained, and shear strengths are measured in one direction with theoretically uniform shear displacement.

2. Improper gripping systems of geosynthetic specimens on the direct shear device affect the accuracy of the results obtained in interface shear testing. If the specimen is not adequately secured to the shearing blocks, it experiences progressive failure and shear strength that deviates from the true field performance of the tested material.

3. A different gripping system can be used to secure the geosynthetic specimens as the ASTM guidelines do not specify the specific type or textured surface to be used. However, the selected gripping surface should be able to provide adequate resistance against slippage of the tested specimens, transfer all applied shear load through the outside surface to the tested specimen, provide a good drainage system if necessary, and should not interfere with the accuracy of the results obtained.

4. The use of proper gripping systems is expected to reduce difficulties encountered during results interpretation and increase the test's result accuracy and reproducibility.

5. Shear displacement rate is an important aspect of interface shear strength testing of geosynthetic materials (i.e. GCL). The ASTM D 5321 recommends that interface tests containing GCL be sheared at a maximum displacement rate of 1 mm/min. If a failure occurs within a hydrated GCL, the test should be repeated using a maximum displacement rate of 0.1 mm/min.

6. The appropriate duration time for full hydration for GCL interface shear tests remains a point of continuing concern. However, most production testing laboratories hydrate GCL for 24 to 48 hours because the little decrease in shear strength is expected if shearing occurs at the same normal stress.

7. Residue strength may not be defined sometimes when a 300 mm x 300 mm direct shear device is used due to its limited in horizontal displacement. Therefore, a reduced shear strength reached beyond the strain softening is called the large displacement (LD) strength.

8. Comparative studies are needed to assess the viability of a multi-interface test in understanding the shear interface in composite liner systems.

3 RESEARCH MATERIALS AND METHODOLOGY

3.1 Introduction

This chapter presents the detailed description of the materials and methods adopted in this research, followed by the experimental procedures.

In order to achieve the objective of the study, the investigation utilized three different classes of geosynthetics that constitute the critical interface component of a lining system in a modern landfill liner i.e. geotextiles, geomembranes, and geosynthetic clay liner. It was necessary to use these materials as their interfaces are the most investigated during the designing of the landfills (Visser 2018).

To ensure that the interface tests were conducted according to internationally recognized standards, a 305 mm x 305 mm direct shear device was selected to measure the interface shear parameters based on the design and testing standards recommended by the Department of Environmental Affairs (DEA) - South Africa (2013); ASTM D5321 (2017) and ASTM D6243 (2018). A summary of all the tests conducted was presented at the end of the section.

3.2 Research Materials

3.2.1 Geotextiles

The bidim A10 (GTX-A) and F-25 SA (GTX-B) geotextiles were used in the study. The material properties provided by the manufacturers are shown in Table 3-1. The 6.4 mm GTX-A was manufactured by Kaytech in Cape Town, Western Cape, South Africa. This product was specifically selected because it was readily available and most importantly, it met the minimum required specification stipulated by the Geosynthetic Institute (GSI) (2013) to be used as a cushion in a landfill. Moreover, bidim is one of the most widely used geotextiles in landfill liners, particularly in South Africa (Stripp 2018), making it suitable to simulate the anticipated field condition.

According to the manufacturer, Kaytech Engineered Fabrics, bidim is a continuous filament non-woven needle punched geotextile, manufactured from 100% recycled polyester coming from discarded cool-drink bottles under controlled conditions (Kaytech Engineered Fabrics Ltd 2015).

The GTX-A used was a dark gray geotextile which had a design life of over 120 years (Kaytech Engineered Fabrics Ltd 2015). When used at a landfill, the bidim A10 protects mostly the HDPE geomembrane liner due to its thickness, high strength and long-term creep (Kaytech Engineered Fabrics Ltd 2015). Additionally, it is easy and quick to install as compared to sand protection layers, especially on slopes.

The 0.7 mm gray, non-woven GTX-B was manufactured in Hammarsdale, KwaZulu Natal, South Africa by Fibertex South Africa. It was selected because it is the most commonly used geotextile-filter in South Africa due to its superior hydraulic properties i.e. permeability and permittivity (Oriokot 2018a).

According to Fibertex, the GTX-B was factory-made from virgin polypropylene fibres with added UV stabiliser. It had uniform material properties i.e. physical, chemical and mechanical. The basic strengths (i.e. tear/tensile) of material were obtained by needle punching the polypropylene fibres, intended to provide strong elastic bonding that resulted in a highly durable and resistant product to all-naturally occurring soil alkalis and acids (Fibertex 2017).

When installed in a landfill application, the GTX-B serves as a filter layer that restricts the movement of solid particles subjected to hydrodynamic forces, while allowing the passage of fluids with little to no increase in pore pressure to the surrounding material; thus, making the structure stable (Fibertex 2017; Oriokot 2018b).

Table 3-1: Properties of the geotextiles (after Kaytech Engineered Fabrics Ltd 2015; Fibertex 2017)

PROPERTIES*	TEST METHODS	UNITS	GTX-A	GTX-B
Mass per unit area	ASTM D5261	g/m^2	1080	140
Thickness**	SANS 9863:13	mm	6.4	0.7
Grab tensile strength	ASTM D4632	kN	4.70	0.58
Grab tensile elongation	ASTM D4632	%	50-80	45-65
Trap. Tear strength	ASTM D4533	kN	2.100	0.17
Punctured (CBR) strength	SANS 12236:13	kN	11.7	1.70
UV resistance	ASTM D4355	%	70	70
Permeability	SANS 11058:13	m/s	0.01	0.07
Pore size, O_{90w}	ASTM D4751	μm	< 75	70
* All values are MARV except UV resistance; it is a minimum value. ** under 2 kPa load				

Microscopic images of the GTX-A and GTX-B were obtained at the Electron Microscope Unit of the University of Cape Town (UCT). A sample was cut at a random section from each standard supplied roll of the non-woven geotextiles and using a Nova NanoSEM 230 microscope, at 200 times magnifications, the fibre arrangement was seen as shown in Figure 3-1.

Figure 3-1 shows that the textiles were a combination of the fibres that were crossed and randomly arranged in multiple directions to mobilize the required strength in all round multi-directional. The GTX-A had continuous filaments which were spun-bonded and needle-punched as it can be seen in Figure 3-1a. The fibres in GTX-B were needle-punched and thermally bonded together as shown in Figure 3-1b.

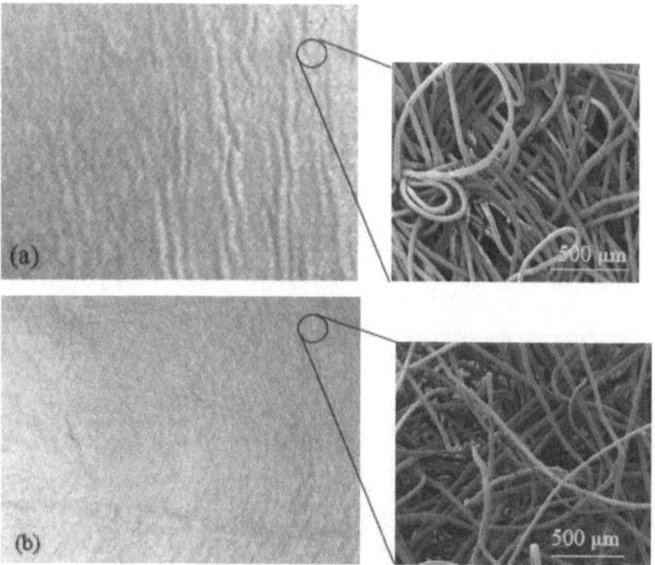

Figure 3-1: Geotextiles used in the investigation; (a) GTX-A and (b) GTX-B

3.2.2   Geomembrane

The geomembrane (GMB) used in the study was a 2 mm black thick double textured, HDPE manufactured by AKS Liner Systems (Pty) Ltd in Cape Town, South Africa. This GMB was specifically selected because it was the most popular geomembrane used in landfills, not only in South Africa (Scheirs 2009; Stripp 2018). The product was manufactured from virgin polymeric resin under controlled conditions, accordingly, the properties were consistent (Stripp 2018).

The GMB had average surface asperity heights of 0.80 mm on one side (side A) and 1.81 mm on the other side (side B). The difference in height asperities was mainly influenced by field

placement, such that side B mobilized more interface shear resistance due to higher asperities (Hardie 2018a). The asperity spacing is shown in Table 3-2 for both sides.

Table 3-2 shows the physical properties measured using a digital diaguage (for asperity height), 100 mm x 100 mm metal sheet (for asperity density) and a rule (for asperity spacing) the by the user. The material specifications obtained from the manufacturer's pamphlet were also included.

Table 3-2: Properties of the GMB and the standards used for determining the specifications.

Properties	Specifications		Units		Standards
Asperity height	Side A		mm	0.80	
	Side B			1.81	
Asperity density	Side A		Per 10000 mm^2	337	
	Side B			211	
Asperity space at center to center	Side A	x-direction	mm	5	
		y-direction		7	
	Side B	x-direction		5	
		y-direction		11	
Asperity direction	Side A		°	90	
	Side B			45	
Density*			g/cm^3	0.946	ASTM D792
Carbon black*			%	2.25	ASTM D4218
Tear resistance*			N	249	ASTM D1004
Puncture resistance*			N	645	ASTM D4833

*(AKS 2018).

In landfill applications, the GMB was mainly installed between the geotextile-cushion i.e. GTX-A and the GCL. It acted as an impermeable layer for contaminate fluids trying to penetrate into the surrounding environment. In Figure 3-2, the asperity surface arrangement of the used geomembrane was shown.

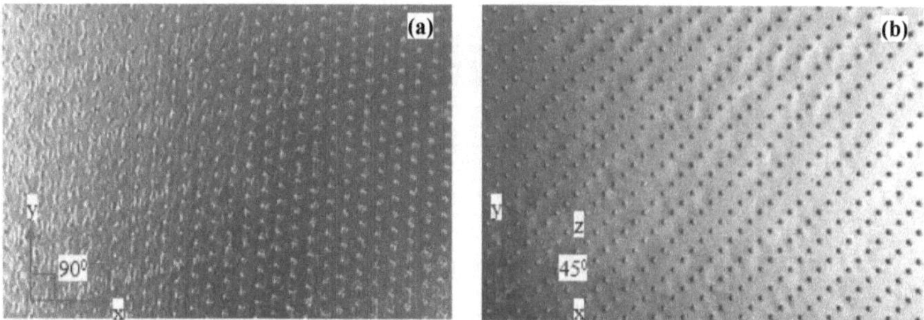

Figure 3-2: Asperity difference of the HDPE used, (a) 0.71 mm and (b) 1.81 mm

3.2.3 Geosynthetic Clay Liners

The application of GCL's as a replacement for the traditional compacted clay in landfill liners has become a common practice nowadays (Stripp 2018). More specifically, the envirofix x800 is the most widely used GCL in South African landfills (Stripp 2018). It was therefore chosen for the evaluation of its interface behaviour against other geosynthetics in this study. Additionally, the envirofix x800 GCL met the recommended GSI guidelines for a GCL to be used in the landfill liner system (Geosynthetic Institute (GSI) 2013).

Envirofix x800 was also manufactured by Kaytech Engineered Fabrics Ltd. The material was typically 2 mm to 2.7 mm thick in its un-hydrated state (Kaytech Engineered Fabrics Ltd 2013). The GCL was a reinforced composite product made up of, from top to bottom, a white polypropylene non-woven geotextile cover, a light brown, dry sodium bentonite powder layer in the middle and a polypropylene slit film woven geotextile carrier layer as shown in Figure 3-3 (Kaytech Engineered Fabrics Ltd 2013).

To ensure reinforcement, polypropylene fibres from the non-woven geotextile were needle-punched through the bentonite and anchored in the woven geotextile with a frictional connection. In Table 3-3 the properties of the GCL as specified from the manufacturer's handbook are shown.

Table 3-3: Properties of the GCL used (Kaytech Engineered Fabrics Ltd 2013).

Properties	Specification	Units	M A R V	Standards
Geotextile Cover Layer	PP nonwoven, white	g/m^2	200	ASTM D5261
Geotextile Carrier Layer	PP slit film, woven	g/m^2	110	ASTM D5261

Continuation of Table 3-3: Properties of the GCL used (Kaytech Engineered Fabrics Ltd 2013).

Properties	Specification	Units	M A R V	Standards
Bentonite Layer (bentonite mass at 0% moisture content)	Quality	Montmorillonite content > 75 %, Sodium Cation Na+ > 60 %		
	Sodium Bentonite Powder	g/m²	3 700	ASTM D5993
	Swell Index (minimum)	mℓ/2 g	≥ 24	ASTM D5890
GCL Mass per Unit Area		g/m²	4 010	ASTM D5993
Grab Strength	MD	N	600	ASTM D4632
	XD	N	600	
CBR Burst	Strength	N	1 400	ISO 12236
	Elongation	%	≥ 15	
Hydraulic Conductivity (maximum)		m/s	$\leq 2.56 \times 10^{-11}$	ASTM D5887
Bonding Process	Needle punched and Thermal Lock™			

PP = Polypropylene   MD = Machine Direction   XD = Cross Direction

M A R V = Minimum average roll value

In landfill applications, the envirofix x800 serves mostly as a barrier for liquid on a base liner or cover system (Visser 2018). It was for this reason as well that the envirofix x800 GCL was selected to be used throughout in the experiments. Figure 3-3 shows the three components of the similar envirofix x800 GCL used in the study.

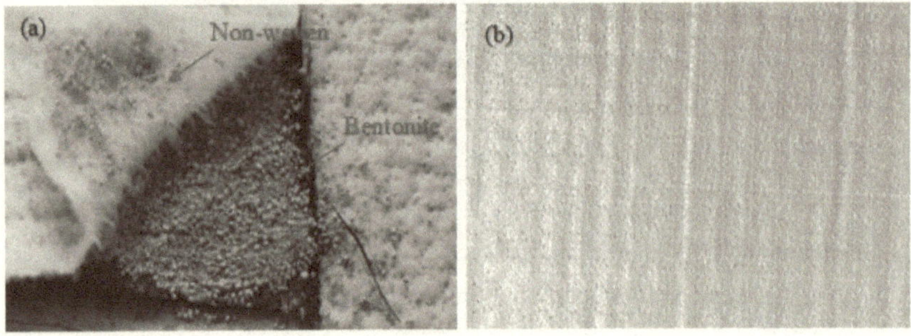

Figure 3-3: Envirofix X800 GCL, (a) non-woven and bentonite – similar GCL (b) woven surface

3.3   Research Testing Equipment

3.3.1   Large Direct Shear

A free-standing automated ShearTrac-III, large direct shear device developed by Geocomp Company in America was used in this investigation. This direct shear equipment consisted of the top (static) and lower (moving) sections. The top shear box measured 305 mm x 305 mm in plan and 100 mm in depth. The bottom shear box had dimensions of 460 mm x 355 mm in plan and 100 mm in depth. Also, the shear device had a water tank (for hydration purposes if required by the test), two high-speed precision micro-stepper motors that control the loading mechanisms for horizontal and vertical load, sensors for measuring load and displacement as well as vertical and horizontal motions embedded controllers for test control and data acquisition (Geocomp 2018). In Figure 3-4, a general configuration of this apparatus is shown.

Figure 3-4: Large ShearTrac-III device (a) Backside (Geocomp 2018) and (b) Front side

According to the manufacturer, Geocomp, the ShearTrac-III had a maximum load capacity of 44.5kN (10,000 lbs) and was capable of applying a constant shear rate of up to 15 mm/minute as well as maximum normal stress of 450 kPa. This ShearTrac-III system generated a fairly homogeneous state of shear stress throughout the specimen, which provides initial stress condition, stress path, and deformation configuration that models numerous field loading conditions more closely than any other test (Geocomp 2018).

The apparatus was mainly chosen in this investigation as it satisfied the minimum requirements described by ASTM-D5321 (2013) standards. Also, a similar equipment had been used by many researchers, including Trauger et al. (1997); Zornberg et al. (2005); Zanzinger & Saathoff (2010); Buthelezi (2017) to determine the interface shear strength between geosynthetics; thus, making it possible to compare the results. Finally, the shear displacement of the direct shear box is theoretically uniform across the width of each tested specimen (Geocomp 2018).

3.3.2 Gripping Systems

Regarding the selection of the gripping system for a particular test, ASTM D5321 and D6243 recommend that a "flat jaw-like" clamping device and a textured gripping surface is normally sufficient. Since the main aim of this study was to investigate the effects of the gripping systems in geosynthetics interface shear strength, two different specimens gripping systems were employed. A 300 mm x 40 mm x 12 mm clamping metal device was used either with the nail plate surfaces or 3M safety-walk slip resistant tape (sandpaper) surface on each shearing block. In this study, the symbols NP and SP represented the nail plate and sandpaper gripping system, respectively.

3.3.2.1 Clamping Device

Two identical, 300 mm x 40 mm x 12 mm stainless steel bars, manufactured together with the ShearTrac-III direct shear apparatus, were utilized as clamping devices in this study. One clamp was used to confine the test specimen at the front end of the bottom shearing block in the direction of shearing. Equally, the other clamp secured another test sample at the rear end of the top static shearing block in the direction of shearing. The clamps were designed in such a way that each test specimen was fastened between the clamping device and the shearing block by the help of five bolts. This assemblage is shown under the specimen preparation section. In Figure 3-5, the detailed description of the clamping device is shown.

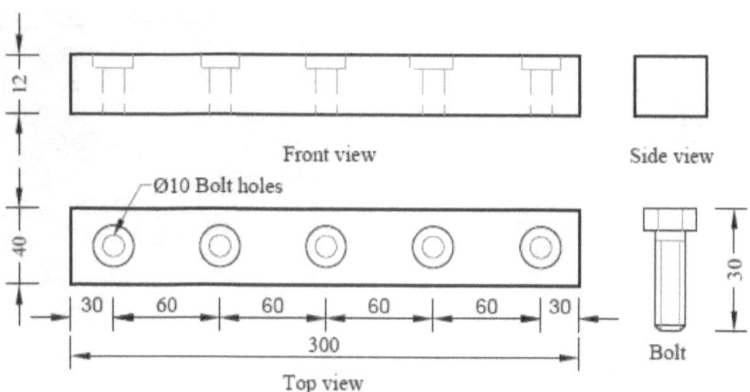

Figure 3-5: Detailed clamping device (all dimensions in mm).

3.3.2.2 Sandpaper

A clear 0.87 mm thick, 3M safety-walk sandpaper was used in the investigation. The product consisted of 399.23 µm to 595.02 µm abrasive particles bonded by a tough, durable polymer to a stable plastic film (of about 1 mm thickness) to form a "coarse" surface (see Fig. 3-6). The bottom surface of the sandpaper was coated with a pressure-sensitive adhesive which was covered by a removable protective brown liner.

The use of the 3M sandpaper was influenced by previous studies (i.e.Fox et al. 1997; Kalumba 1998) who have utilized similar sandpapers. Also, the 3M sandpaper can be installed to provide resistance against slips in dry, wet, or oily conditions, thus, making it suitable to be used even in tests conducted under saturation condition. Furthermore, the sandpaper exceeded the recommended standards for slip resistance in accordance with ASTM requirements (United States Dept. of Labor 2012).

An SEM image taken at 100 times magnification is shown in Figure 3-6. The microphotography shows that the sandpaper mainly consisted of irregular angular shape particles with a rough surface. These particles were randomly bonded to plastic film and had size particles that ranged from about 399 µm to 595 µm.

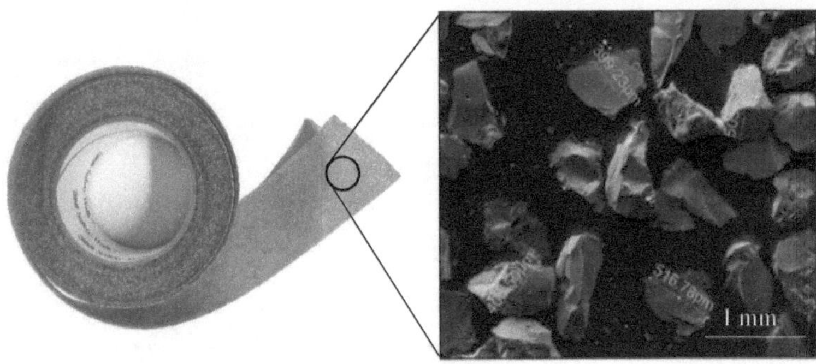

Figure 3-6: 3M safety-walk slip sandpaper used in the study

3.3.2.3   Nail Plate

Based on previous studies (i.e. Zanzinger & Saathoff 2010; Zanzinger & Alexiew 2000; Buthelezi 2017), two nail plates were specifically designed in the UCT civil engineering workshop for gripping the geosynthetics in the investigation. The plates were machined from a single piece of aluminium sheet, hence, they had high strength (> 75N) and durability. Another reason for choosing the aluminium was due to its ability to not subject to deterioration in the wet environment, thus, can be used in tests conducted under saturation condition.

The top and bottom plate were trimmed to a dimension of 303 mm x 303 mm x 4 mm and 480 mm x 303 mm x 4 mm respectively. On each plate surface, 3 mm diameter holes were drilled at an equal spacing of 13 mm (169 mm^2). However, it is necessary to note that 80 mm on one side (length direction) of the bottom plate was left out without the holes. This space was used to drill five through bolt holes of 10 mm in diameter, spaced equally at 60 mm (from the nails) and 20 mm from the edge on the bottom plate surface in line with the existing five holes on the bottom shearing block (see Fig.3-7). This was to enable the plate to be fastened tightly to the shearing block prior to shearing to provide a better resistance to slippage.

Sharp stainless-steel nails of 6 mm in height and about 3.6 mm in diameter were screwed vertically into the pre-drilled holes (3 mm holes), thus, leaving approximately 2 mm height nail pointers protruding on the topside of each plate as recommended by ASTM D 6243 (Soleimanian et al. 2016). It was this topside of each gripping plate that was in contact with the test samples. Overall, the top and the bottom gripping plate had 552 and 768 nails on the surface, respectively.

It was assumed that the selected height and spacing would be capable of providing the needed resistance against slippage. Moreover, previous researchers have used the same height and spacing in their studies, hence, it would be possible to compare the results.

Also, a 3 mm diameter drainage hole was drilled at each nail slot intersection (about 13 mm or centre) on both surfaces to allow water to move through the plate from the slotted side to the gripping side. The drainage holes provided free access to water during hydration, consolidation, and shearing. It was anticipated that a 3 mm diameter hole was capable of providing sufficient drainage at each 169 mm^2 area occurred on the plate. Moreover, the nail and drainage spacing was influenced by machine capability.

In general, the plates were designed in such a way that during the shearing process, the two nail plate teeth may not touch each other at any time. This was important in transferring the applied 'total' normal stresses over the total area of both the upper and lower side of the tested specimen. Figure 3-7 shows the details of the nail plates used in the study.

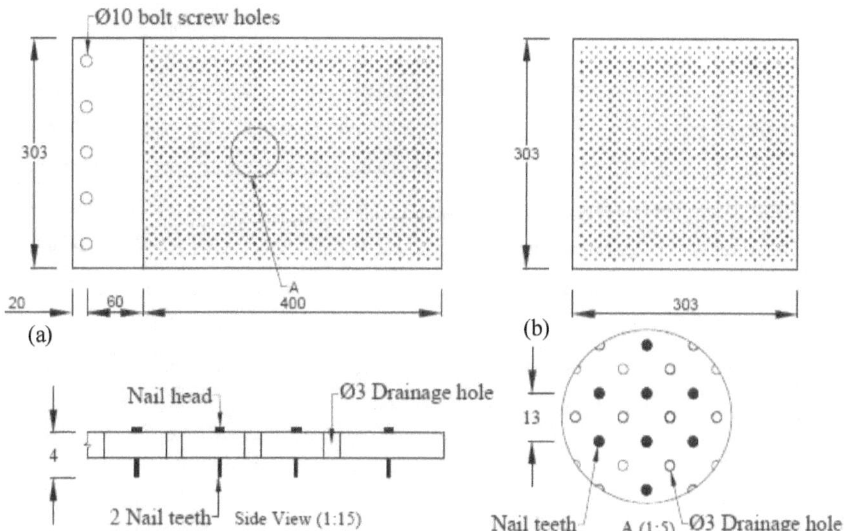

Figure 3-7: Detailed design of the nail plates: (a) Bottom and (b) Top (all dimensions in mm).

3.4   Specimen Preparation

Geosynthetic specimen preparation in each category under investigation was conducted in accordance with the guidelines specified in ASTM D5321 (2017) except for GCLs which followed the ASTM D6243 (2018) standards. The test samples in each test were also prepared with care to obtain similar testing conditions throughout all the experiments.

As described in ASTM D5321, a metal substrate that served as a replacement for soil material was gently placed inside the bottom shearing box. This substrate was supplied together with the direct shear apparatus, hence, it was specifically designed for such a direct shear box. Moreover, previous researchers, including Buthelezi (2017) reported that the metal substrate, particularly the one used in this study increased the accuracy and reproducibility of test results. The substrate had dimensions of 400 mm in length, 300 mm in width and 100 mm in height.

Once the substrate was inside the shear bottom box, a gripping surface (either nail plate or sandpaper) was then carefully placed on top of the respective substrate. This was to simulate the friction characteristics between two geosynthetics or geosynthetic and the adjust material in the field (Kalumba 1998).

In the tests that involved the use of the sandpaper as a gripping surface, an aluminium dead weight of mass 3.3 kg with dimensions of 300 mm x 300 mm x 12 mm was used as a substrate for the top shearing block. Thus, the 50 mm width sandpaper was carefully cut from the respective industrial supplied roll and glued on both the bottom and top metal substrate surface. The sandpaper pieces were glued closely to each other on the topside surface (surface in contact with test samples) of the substrate and left for about 15 minutes to cure (without placing anything on top) in order to form a gripping surface as shown in Figure 3-8. The selection of this aluminium plate was influenced by its ability to resist corrosion, making it suitable to use in tests performed in wet conditions.

Figure 3-8: Sandpaper top gripping surface

Geosynthetic test samples were trimmed parallel to the factory roll direction (machine direction) from their respective industrial supplied geosynthetic rolls to fit either the top or bottom shearing blocks. The machine direction plan was preferred as geosynthetics are mainly placed in the field with the machine direction matching to the slope (Hardie 2018a; Oriokot 2018a; Stripp 2018). It was, therefore, necessary to simulate the site anticipated conditions during the experiments. Moreover, geosynthetic properties are sometimes direction dependent, hence, the direction in which they are tested can influence the values of certain properties (Sarsby 2007).

The test samples for each experiment were cut at random sections from the respective supplied geosynthetic rolls. However, it is imported to mention that each test specimen was carefully inspected to ensure that no damaged samples were tested.

Table 3-4 shows the specimen configuration followed in the investigation. The sizes of the test samples shown in Table were sufficient to cover the entire top and/or the bottom shearing surface of the shear box such that no area correction was required (ASTM D5321 2017; ASTM D6243 2018). Also, the respective specimen dimensions enabled the facilitation of the edge clamping of the test samples to the end of the shearing blocks as recommended by ASTM D5321 (2017) and ASTM D6243 (2018), except for the middle test specimens used in the multi-interface tests.

Table 3-4: Specimen configuration

Specimen configuration	Size of Test specimens*	Interface Investigated		
		GTX-A/GMB	GMB/GCL	GCL/GTX-B
Top shearing block	305 mm x 325 mm	GTX-A	GMB	GCL
Bottom shearing block	305 mm x 520 mm	GMB	GCL	GTX-B
Bottom shearing block	305 mm x 520 mm	GTX-B		

*Cutting accuracy of ±0.5 mm

Continuation of Table 3-4: Specimen configuration

Specimen configuration	Size of Test specimens*	Interface Investigated GTX-A/GMB/GCL/GTX-B
Top shearing block	305 mm x 325 mm	GTX-A
Middle	305 mm x 325 mm	GMB and GCL
Bottom shearing block	305 mm x 520 mm	GTX-B

*Cutting accuracy of ±0.5 mm

Figure 3-9 shows the tools and equipment used in preparing the test specimens. The GMB test samples were cut using the mechanical saw machine (due to strong fabric) from the civil workshop while each GCL and GTX-A/B specimen was cut using a pair of scissors (due to soft fabric). Once all the test samples were cut into their respective size, a 10 mm diameter mechanical-hole puncher and a hummer were used to make five gripping holes with an accuracy of ±0.5 mm at one edge of each trimmed test specimen. The holes were 20 mm from the edge and spaced at 60 mm to match the holes on the shearing blocks and clamping device (see Fig. 3-9g). These holes were used to secure the test sample to the shearing blocks using the clamping device to provide extra strength to hold the tested samples in position and enforce failure of the specimen to occur at the desired interface (Fox et al. 2004).

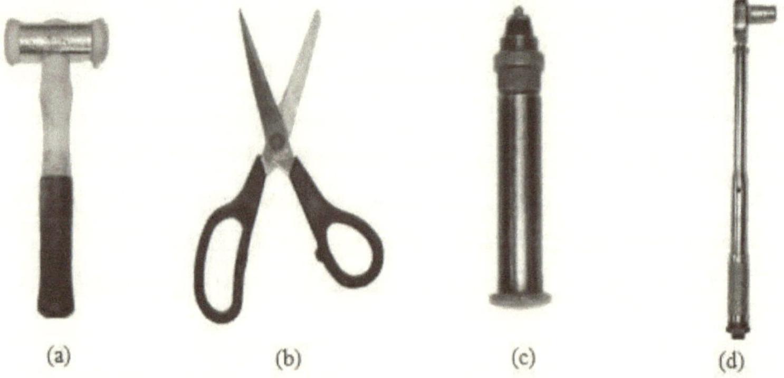

Figure 3-9: Tools and equipment used in specimen preparations: (a) hammer, (b) pair of scissors, (c) mechanical puncher hole and (d) ratchet bolt driver.

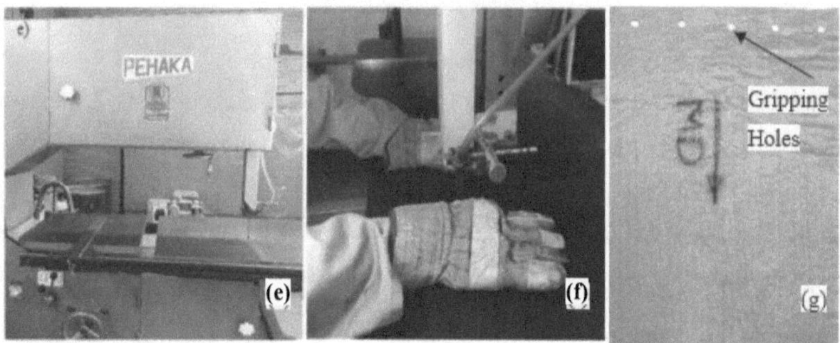

Continuation of Figure 3-9: Tools and equipment used in specimen preparations; (e) mechanical saw machine (f) cutting the test specimens and (g) test specimen.

3.4.1 Single Interface Test

In each single interface test category, the trimmed top and bottom samples were carefully fixed at one end, i.e. the edge with holes, of the respective shear block with the help of the clamping device and five bolts (see Fig. 3-10). The test samples were fitted in a slightly but carefully stretched mode with the shearing direction for each test specimen matching its machine direction. Figure 3-10 shows the top and bottom test specimens secured to the respective shearing blocks.

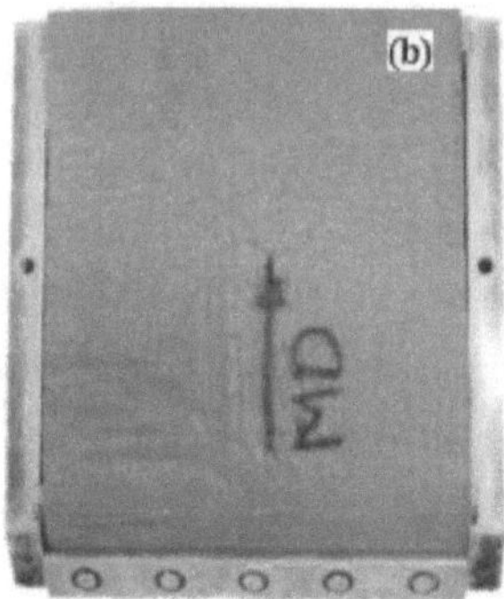

Figure 3-10: Test specimens fitted to the shearing block, (a) Top and (b) Bottom

The upper shearing block together with the clamped test specimen was then carefully held and placed on top of the bottom shearing block as illustrated in Figure 3-11. Using two alignment screws, the top shearing box was gently aligned and held in position with the bottom shearing block to prevent any movement during handling. The upper gripping surface was then placed inside the shearing block with care, such that the topside of the gripping surface was in contact with the upper test specimen (see Fig. 3-11). Figure 3-11 shows the cross-section of the set up.

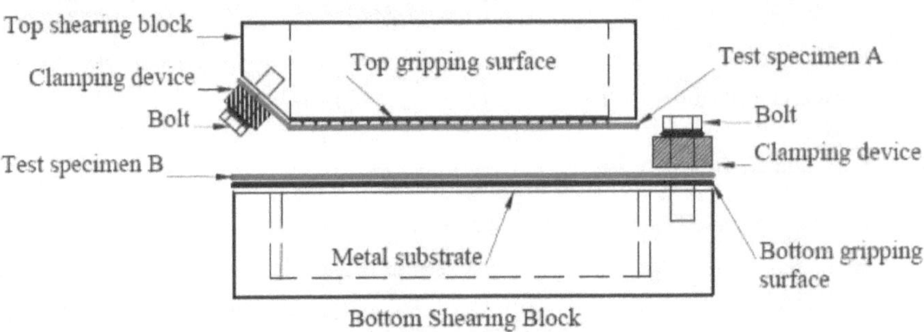

Figure 3-11: Test sample configuration for single interface test

3.4.2   Multi-interface Test

In multi-interface test configurations, the same procedure followed in the single interface test preparation was adopted, but two additional test samples were added to simulate the anticipated field condition as explained earlier in section 2.3.3. The GTX-A and GTX-B were respectively fixed with care to the top and bottom halves of the shearing block with the help of the clamping device and bolts (five on each clamping device). In a gently but carefully stretched mode, the GMB and then the GCL was placed on top of the GTX-B test sample without being clamped, such that the GMB and GCL were in between the GTX-A and GTX-B test specimen as illustrated in Figure 3-12. This was to allow failure to occur at the weakest interface during interface shearing as highlighted in section 2.3.3.

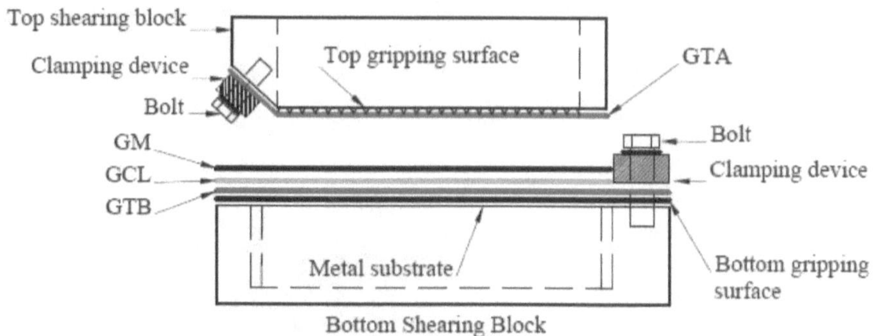

Figure 3-12: Test sample configuration for multi-interface test

3.4.3   Final Assemblage of Apparatus

At this stage, in either single or multi-interface test preparation, a 10 kg loading plate of dimensions 300 mm x 300 mm x 22 mm was carefully placed over the top gripping surface inside the upper shearing block as described by the direct shear manufacturer. The loading plate, whose primary purpose was to distribute the applied normal force to the test samples in a shear box, was designed and supplied by Geocomp together with the ShearTrac-III, direct shear apparatus. Additionally, a 79.1 g steel ball-bearing with a diameter of 26.6 mm was then gently placed in the spherical seating on the loading plate as specified by Geocomp (2018). Figure 3-13 illustrates the assemblage.

The whole assemblage was then carefully loaded onto the low-friction machine testing bed by sliding it into the ShearTrac-III base container. Using the connected desktop computer, the assemblage was adjusted by moving it horizontally such that the steel ball-bearing was aligned to the vertical loading device of the apparatus as shown in Figure 3-14. The assemblage was then bolted in position by two steel reaction beams that span on top of the test chamber and are screwed to the frame of the machine (see Fig. 3-13). The beams provided the reaction force to the top shearing box to ensure that only the bottom shearing block moved during shearing.

Once everything was in order, the two aligning screws holding the top shearing block to the bottom shearing block were carefully removed with conscious of not causing any displacement between the two shearing blocks.

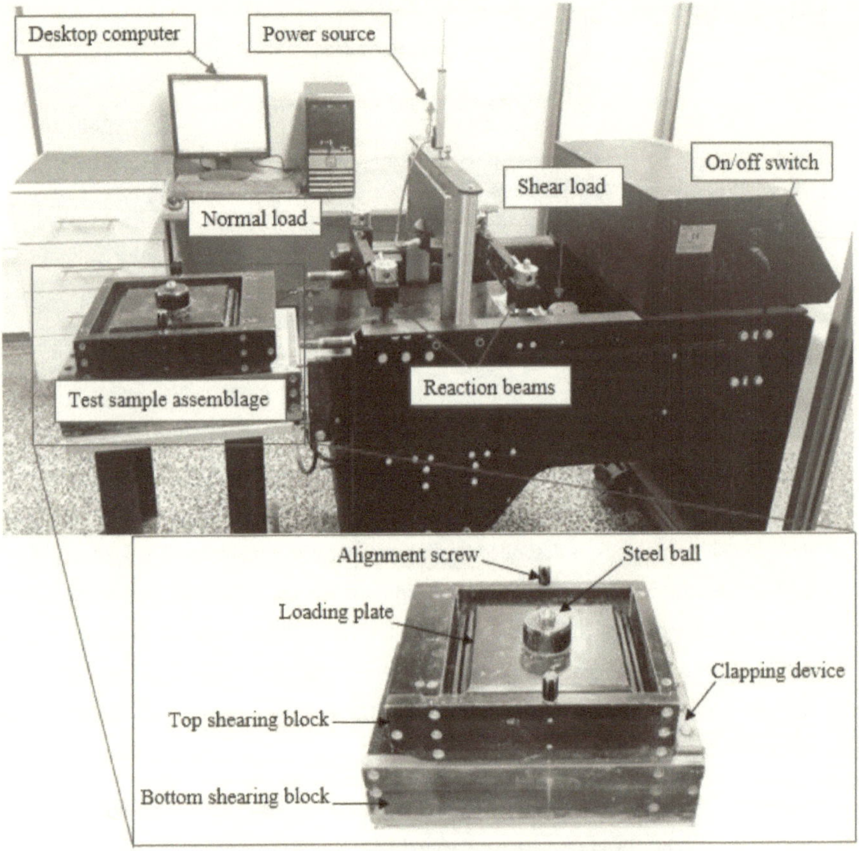

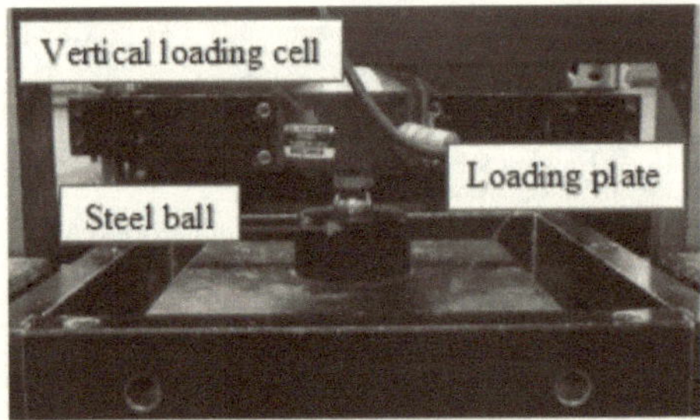

Figure 3-13: Assembly of apparatus

Figure 3-14: Vertical loading cell coinciding with the steel ball on the loading plate of the direct shear apparatus.

3.4.4   Hydration

In landfill composite liner systems, GCLs are expected to hydrate soon after installation due to the liquids from the stored waste and/or the rains. Hence, it is essentially important to hydrate the GCLs under the expected field condition when determining their interface shear behaviour (Stripp 2018).

In this study, all the GCL specimens were hydrated under a normal stress of 17 kPa for 24 hours. The 17 kPa was selected to simulate the possible initial load that would be imposed on the lining system before waste disposal as reported by other researchers, including Eid and Stark (1997); Fox and Stark (2004). Moreover, it was recommended to hydrate the GCL at low normal stress as it resulted in more water being absorbed into the double layers of the bentonite particles, thereby, presenting the worst case scenario (Fox & Stark 2004).

The saturation time of twenty-four (24) hours employed in this study for all interface tests involving GCL was also influenced by previous studies. This hydration time had been commonly used and found to be sufficient for a sound conclusion to be drawn (Fox & Stark 2015; McCartney et al. 2009; Fox et al. 2004). Conversely, a hydration time of one hour was used instead of 24 hours for the GMB/GTX-A and GMB/GTX-B interface tests because no soils (i.e. Clay in the GCL) had to undergo consolidation (Stark et al. 2015). The use of one hour was to allow enough time for the geomembrane and geotextile to engage prior to shearing (Stark et al. 2015).

3.5   Test Procedure

Once the setup of the direct shear apparatus and all testing samples were placed in order, the ShearTrac-III, custom software used to run the test was opened on the connected desktop computer and testing parameters were entered manually. These parameters including, the normal stress, shear rate, horizontal displacement, and consolidation time.

According to the ASTM D5349 and D6243, at least three normal stresses are needed to develop a shear strength envelope for geosynthetics. Fox & Stark (2004), however, recommended that the normal stresses be carefully selected over a wide range due to the nonlinearity of shear strength envelopes for geosynthetics (i.e. GCLs). Therefore, five normal stresses were selected to represent the range of stresses expected in a landfill having a unit weight of 9.81 kN/m^3 and a

maximum height of 30 m as described by Visser (2018). The normal stresses were calculated using equation 3-1, hence, the tests were performed at normal stresses of 50, 100, 200, 295, and 400 kPa to represent the varying load conditions experienced by the liner system throughout the design life of the landfill (Visser 2018). The lowest normal stress, i.e. 50 kPa was used to represent the stress-dependent nature of the strength envelopes at lower effective stresses (Stark & Choi 2004).

$$\sigma_n = \gamma \times h$$

Equation 3-1

Where σ_n = normal stress

γ = maximum unit weight of the waste

h = maximum height of the landfill

A constant shearing rate of 1.0 mm/min was used in all the interface shear tests which did not involve the GCL specimens as no excess pore pressures were anticipated on the interface (ASTM D6243 2018). Conversely, a recommended shear rate of 0.1 mm/min was used in all interface tests which involved GCL test samples (ASTM D6243 2018). This shearing rate was relatively slow to ensure insignificant buildup of excess pore pressures at failure (ASTM D6243 2018).

The ASTM D5321/6243 standards recommend a minimum shearing displacement of 50 mm in the geosynthetic interface shear test. In this study, a shearing displacement of 70 mm was used throughout all performed experiments. Moreover, the direct shear device used for the tests had a maximum displacement of 75 mm but for safety reasons, the 70 mm was employed. Therefore, the resulting 'residual' strengths were referred to as large displacement (LD) shear strength as explained in section 2.3.2.

A 30 minute consolidation time was adopted from previous studies (i.e. Fox & Stark 2004) and used throughout all the experiments. It was anticipated that this consolidation time was sufficient for the gripping surfaces to fully engage with the test samples and also allow the excess pore pressures to essentially reduce to zero prior to the start of shearing.

The project details describing the test conducted were also manually entered to keep track of each different test conducted. The shear device was then calibrated according to the manufacturer's manual before shearing was initiated to ensure uniformity of the results.

Once the test was initiated, the test samples were allowed to pre-consolidate until the time for 100 per cent primary consolidation (t_{100}) was achieved. Tap water drawn from the UCT geotechnical laboratory (about 15 littles) was carefully poured into the apparatus water tank until all test specimens were fully submerged for hydration purposes. The tap water was used for its mineralogy composition was not considered to affect the interface shear strength behaviour of the tested geosynthetics (Thiel & Criley 2005; CETCO 2014; Buthelezi 2017; Department of Water Affairs and Forestry 1996). The saturation of test samples represented the least favourable condition of hydration that could be expected in a landfill (i.e. during rainy season conditions).

Upon completion of the required hydration time, a gap of approximately 5 mm to 10 mm, between the upper shearing block and the lower shearing block, was created to prevent friction between the two shear blocks during shearing (ASTM D5321 2017). At this stage, the test samples were ready to be sheared.

In order to initiate shearing, a shear force calculated by the direct shear program was applied to the bottom shearing block from the motor apparatus through a lead screw of diameter 37.5 mm to produce a shear stress. This resulted in the bottom section of the shearing box, moving in relation to the top static shearing block at a user-specified shear rate i.e. 1 mm/min.

During the shearing phase, the ShearTrac-III software program continuously recorded all the data sensors, made necessary corrections to maintain shear rate, graphically displayed the test data on the computer screen, and captured the data in an output file. All the tests automatically ended when the horizontal displacement of 70 mm was reached, as explained earlier.

3.6 Testing Program

A series of forty direct shear tests were performed using the two different gripping systems. On each gripping system, twenty tests were conducted at five different normal stresses, namely 50, 100, 200, 295 and 400kPa. Five of the tests for each gripping system were tested using a multi-interface test method. Table 3-5 presents the complete test program performed and the key factors considered which have important effects on the results of direct shear tests.

Table 3-5: Testing scheduled per gripping system

Interface Test Configuration	GCL Hydration		Shearing rate (mm/min)	Normal Stress (kPa)
	Normal Stress (kPa)	Saturation Time (hr.)		
SINGLE INTERFACE				
GTX-A/GMB	Respective normal stresses for testing	1	1.0	50
				100
				200
				295
				400
GMB/GCL	17	24	0.1	50
				100
				200
				295
				400
GCL/GTX-B	17	24	0.1	50
				100
				200
				295
				400
MULTI-INTERFACE				
GTX-A/GMB/GCL/GTX-B	17	24	0.1	50
				100
				200
				295
				400

3.7 Data Processing

3.7.1 Output Processing

The ShearTrac-III, large direct shear device through the connected desktop computer processed all the data i.e. the tables and figures for shear stress versus horizontal displacement and shear stress versus applied normal stress. Upon completion of each tested interface, the recorded data were exported into a created template file in Microsoft Excel software (MEs) and shear strength graphs such as shear stress versus horizontal displacement and shear strength versus normal stress were plotted. For any applied normal stress, the shear strength data points were connected by a best line(s) or curve generated within the MEs to determine the values of the interface friction angle and adhesion. The secant value of the gradient of the line or curve yielded the interface friction angle while the intercept on the vertical (shear stress) axis gave the adhesion value.

3.7.2 Failure Envelop Criterion

The failure envelopes plotted in the MEs displayed three types of graphs, namely linear, bilinear and curvilinear. The linear failure envelopes described a proportional relationship between the applied normal stresses and shear strengths (i.e. peak or LD). These envelopes were modelled as a function of the applied normal stress using the Mohr-Coulomb equation 3-2 as recommended by ASTM D5321/6243 standards.

$$\tau = \alpha + \sigma_n \tan \phi$$

Equation 3-2

Where: $\tau =$ Peak or LD shear strength

$\alpha =$ Adhesion corresponded to the inclination of the vertical axis

$\sigma_n =$ Applied normal stress

$\phi =$ Interface frictional angle corresponded to the inclination of horizontal axis

Also, equation 3-2 was used in bilinear failure responses to define the failure envelope. The equation was fitted twice between two separate smaller shear stress ranges. This is because the shear strengths in bilinear envelopes increased with increasing normal stress until reaching a certain normal stress teamed, critical stress (Buthelezi 2017). Beyond this critical stress, the rate of shear strengths suddenly changed, concave downwards, with increasing applied normal stress forming a bilinear curve. This caused an abrupt change in interface friction angle and adhesion, thus, two interface friction angles and adhesion were determined.

The curvilinear failure envelope demonstrated that the shear strength of the tested samples was dependent on the applied normal stress and that the shear stress tangent changed with the horizontal displacement. To model this type of failure behaviour, an attempt to use a numerical technique proposed by Giroud et al. (1993) was made, butt unrealistic values were obtained. Therefore, the nonlinearity behaviour of the interface shear strength was estimated by a hyperbolic equation in the form of equation 3-3. This approach was in line with the studies conducted by other researcher, including Esterhuizen et al. (2001), Jogi (2005), KavazanjianJr. et al. (2012) and Bacas et al. (2015) who observed the nonlinearity behaviour of the geosynthetic interfaces on a shear stress versus normal stress graph. Moreover, the constant coefficients can be easily generated within the MEs as compared to assuming, i.e. visually, some input constants in the equation proposed by Giroud et al. (1993).

$$\tau = a\sigma_n^2 + b\sigma_n + c_a$$

Equation 3-3

Where: τ = Peak or LD shear strength

c_a = Adhesion corresponded to the inclination of the vertical axis

σ_n = Applied normal stress

a, b and c = coefficients obtained through a curve fitting technique with the experimental data

Furthermore, the equation was preferred as the coefficients constants were found with good fits $\left(0.99 \leq R^2 \leq 1\right)$ for all interfaces presented.

3.8 Quality Assurance

3.8.1 Measures Implemented

To ensure that all tests were performed up to standard, the following measures were implemented in each test conducted:

- In each experiment, new geosynthetic test specimens with no damage were trimmed to the same respective dimension in accordance with the machine direction of the industrial supplied geosynthetic rolls.
- For all interface tests conducted, the test sample were tested with the respective surface matching the site installation condition as illustrated in Table 3-5.
- Freshwater drawn from the laboratory tap was used in each test for hydration after thoroughly cleaning the apparatus water tank to ensure no contamination of test samples.
- The equipment used throughout the study was calibrated as recommended by the direct shear manufacturer and ASTM D5321/6243 standards.
- All the tests were conducted under similar physical and climatic condition (air-conditioned laboratory at 22 °C and humidity of between 50-70%) as recommended by the ASTM D5321/6243 standards.

3.8.2 Repeatability Test Results

The testing procedures and results in this study were verified for reproducibility by conducting one replicate test from a single interface (first test performed) and one of the multi-interface test. These repeated experiments were conducted at the same normal stresses considered in the investigation (50, 100, 200, 295 and 400 kPa). The resultant shear stress responses are shown in Figure 3-15.

In analysing the repeatability results presented in Figure 3-15, a percentage difference in the mobilized maximum and large displacement (LD) strength was calculated using equation 3-4 (Covemaeker 2017). It was assumed that if the respective percentage differences were less than 10%, then this would be evidence of repeatability (Ernest 2014).

$$PD = \left| \frac{A-B}{A} \right| \times 100$$

Equation 3-4

Where PD = Percentage difference

A = First value

B = Second value

In Figure 3-15, the shear stress behaviour for the repeated tests are presented. It was evident from Figure 3-15, that at any particular normal stress, the plots displayed a similar shear stress relationships response for each interface considered. The shear strength increased with increasing normal stress to reach the maximum (peak) shear strength at a certain horizontal displacement. Beyond the maximum shear stress, the shear stress dropped to approach a large displacement (LD) strength in each interface tested (see Fig. 3-15).

In Table 3-6, the maximum and LD shear stresses read-off from Figure 3-15 are presented for comparisons. It was noticed, in Table 3-6, that the percentage difference in the achieved maximum and LD strengths for the repeated tests was less than 9%, which is within the acceptable limit considered in this investigation (Ernest 2014). Based on these results, it was, therefore, concluded that the procedures adopted in this study and the results obtained are reproducible.

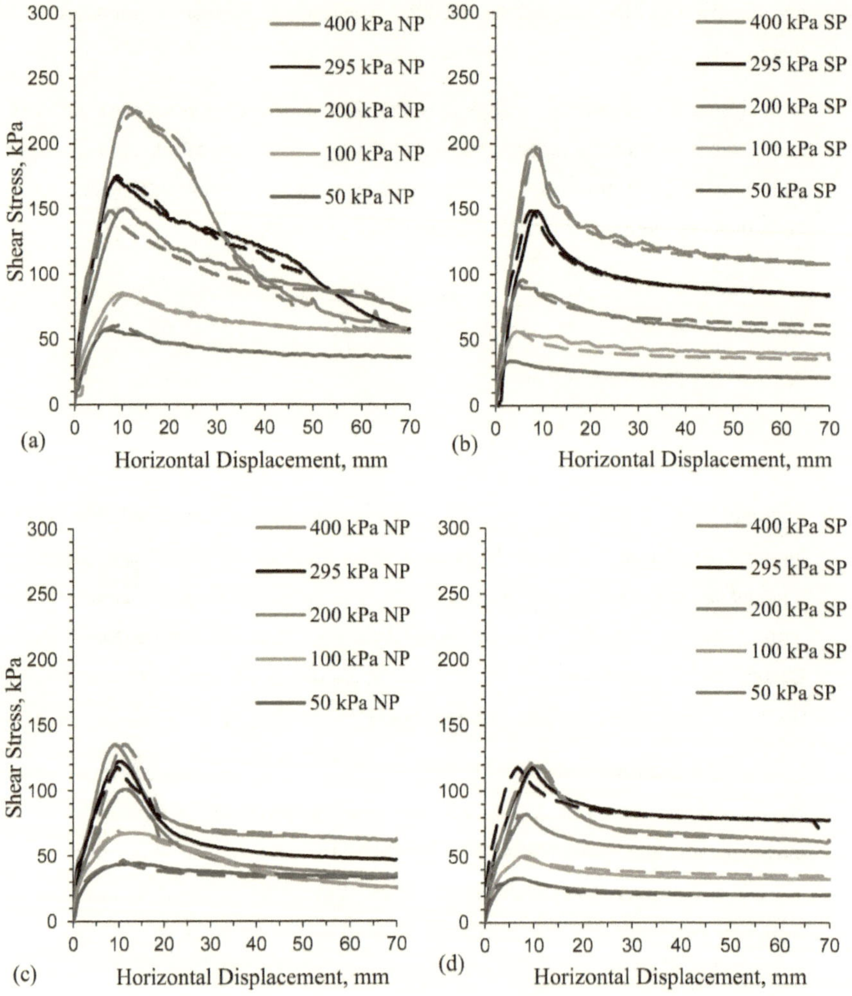

Figure 3-15: Shear Stress Responses; (a) – (b) Single Interface Test (c) – (d) Multi-interface Test

Table 3-6: The percentage difference in the repeated test results.

Interface Combination	Normal Stress	Units	Percentage Difference			
			NP		SP	
			Peak	LD	Peak	LD
Single Interface	50	%	6.1	0.8	8.0	2.3
	100		1.9	1.3	2.1	5.4
	200		1.3	1.8	4.5	2.9
	295		1.1	0.2	2.0	0.1
	400		3.0	0.2	1.5	1.9
Multi-Interface	50		4.5	1.0	6.3	5.0
	100		2.2	2.7	1.7	3.5
	200		1.9	0.0	1.8	0.9
	295		4.0	0.0	2.5	0.6
	400		0.7	0.3	2.5	3.1

4 RESULTS, ANALYSIS AND DISCUSSION

4.1 Introduction

This chapter presents in detail the results of all direct shear tests conducted at the geosynthetic/geosynthetic interface using the two gripping systems, followed by a discussion.

The chapter was divided into four main sections. The measured interface shear responses were presented and discussed in the first section based on the gripping systems employed. The second section tackled the failure envelopes for both methods of gripping systems, while the third and last sections focused on the failure envelope and the critical interface based on the research findings, respectively.

4.2 Shear Stress versus Horizontal Displacement

4.2.1 Introduction

The development of all the shear stress versus horizontal displacement relationships for both single and multi-interface shear tests performed in the study are presented in this section. For each interface test considered, a separate graph was produced. The interface tests were conducted at five different normal stresses i.e. 50, 100, 200, 295 and 400 kPa, thus, there are five plots of the same type of line in each graph. The two gripping systems, nail plate and sandpaper, utilised in the study were differentiated by the symbols NP and SP as shown in the legend, respectively.

4.2.2 Geotextile-cushion/Geomembrane (GTX-A/GMB) Interface Test

The geotextile-cushion (GTX-A) used in the test, with a thickness of 6.4 mm, was a continuous filament non-woven needle-punched geotextile, made from 100% recycled polyester coming from discarded cool-drink bottles (Kaytech Engineered Fabrics Ltd 2015). The geomembrane (GMB) on the other hand, was a 2 mm thick double textured, manufactured from the virgin polymeric resin (AKS 2018). This GMB had an average surface asperity height of 0.80 mm on one side (side tested against GTX-A) and 1.81 mm on the other side.

Relationships for direct shear stress versus horizontal displacement for GTX-A/GMB interface shear tests are presented in Figure 4-1. The interface shear tests were performed after one hour of test samples been fully submerged in water at their respective shearing load i.e. 50, 100, 200,

295 and 400 kPa. A recommended constant shearing rate of 1.0 mm/min was used during each experiment as no excess pore pressures were expected at the interfaces (ASTM D6243 2018).

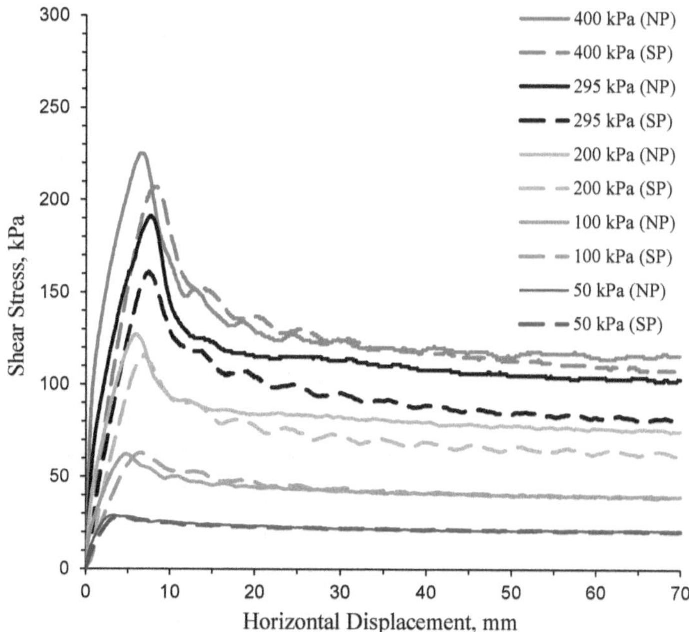

Figure 4-1: Shear stress versus horizontal displacement relationships for GTX-A/GMB interface.

The shear strength curves, as observed in Figure 4-1, followed a theoretical pattern of the shear stress versus horizontal displacement relationship for geosynthetics, i.e. the interface shear stress development increased with corresponding horizontal displacement until reaching the maximum (peak) interface shear strength. Beyond the peak strength, a clear shear strength reduction with further horizontal displacement was observed (see Fig. 4-1). These shear responses followed similar treads to the corresponding findings of other researchers (i.e. Bacas et al. 2011; Bacas et al. 2015; Buthelezi 2017).

In all the interface shear tests performed, the shear stress versus horizontal displacement graphs exhibited an initial rapid increase in the shear stresses to reach the peak strengths, regardless of the gripping system utilized. This stage of shear stress development was classified as a pre-peak and as seen in Figure 4-1, the relationships were all non-linear throughout the tested interfaces.

It was, however, noticed that for any particular pre-peak stress, the NP achieved the shear strengths at a lower horizontal displacement as compared to the SP. This may have been caused by an early engagement of the NP with the test samples, i.e. the nail plate teeth 'punching' into the fabric of the GTX-A test sample as well as forming an intimate engagement with the GMB test specimen, thus, mobilizing a strong interlocking bond during shearing (Allen & Fox 2007). Moreover, the ASTM D5321 (2013) states that the variations in the horizontal displacement at respective achieved shear strengths are due to the difference in test specimen engagement with the gripping systems used during the experiments.

Following the pre-peak strengths, the shear stress curves reached the maximum strengths for all the normal stresses considered (see Fig 4-1). The peak strength, according to Kalumba (1998), can also be defined as the failure point of the tested material (i.e. geosynthetics). In this study, Kalumba's (1998) definition was, therefore, adopted. Comparing the peak strength magnitude in Figure 4-1, it was observed that the failure points were all 'well-defined' and proportional to the increasing normal stresses irrespective of the gripping system. In fact, both gripping systems, NP and SP, mobilized the same peak strength of 29 kPa and 63 kPa at lower normal stresses of 50 kPa and 100 kPa, respectively. This implied that the characteristics of the gripping systems did not show any effect on the measured maximum strengths as the values were very much the same. At higher normal stresses (i.e. 200, 295 and 400 kPa), however, it was observed in Figure 4-1, that as the applied normal stresses increased, diversions in the mobilized maximum shear strengths between the two gripping systems were identified. These differences in the measured peak strength can be seen clearer in Table 4-1.

It is evident, as detailed in Table 4-1, that the difference in determined peak strength increased with increasing normal stresses. For instance, at normal stresses of 200, 295 and 400 kPa, the peak strengths mobilized by NP ranged from 11-30 kPa higher than that of the SP, with 11 kPa recorded at a normal stress of 200 kPa.

Beyond the failure point, the shear strengths of the tested samples decreased rapidly to remain constant at the LD strengths. During this stage, it was considered that the test samples were going through a plastic deformation state, a process referred to as strain softening (Buthelezi 2017). It was observed in Figure 4-1 that the higher the applied normal stress, the higher the strain softening became. This strain softening behaviour was associated with the dislocation movements generated within the crystal structure of the tested specimens (Jayasuriya 2017).

It was also observed in Figure 4-1 that at any applied normal stress (50, 100, 200, 295 & 400 kPa), the shear stress plots approached a steady-state interface shear strength value termed large displacement (LD) after a certain horizontal displacement, depending on the gripping system used. The LD strengths, which are not residual strengths (Triplett & Fox 2001), remained stable under further horizontal displacement until reaching a horizontal displacement of 70 mm to end the experiment (see Fig. 4-1). As in the case of the peak strengths, the LD strength values mobilized by each gripping system we read-off from Figure 4-1 with their respective applied normal and included in Table 4-1. It can be seen from Table 4-1 that both gripping methods employed approached the same LD strengths at lower normal stresses, 29 kPa and 21 kPa for 50 kPa and 100 kPa, respectively. At higher normal stresses, however, the NP attained higher LD strengths as compared to the SP. The LD strength reached by the NP, as seen in Table 4-1, were 13.3, 21.6 and 8 kPa higher than that of the SP for the tests conducted at 200, 295 kPa and 400 kPa, respectively.

Owing to the influence of the gripping system, particularly the surface texturing (teeth), the difference in the shear strengths (peak and LD) was attributed to the gripping surface engagement with the tested specimens. It was anticipated that the higher peak and LD strengths obtained by the NP were due to the intimacy engagement with the tested geosynthetic specimens (Allen & Fox 2007). This created a strong interlocking bond which offered higher shear resistance between the test samples and the gripping surface, thus, transferring the applied normal stresses onto the tested specimens (Fox et al. 2004).

Having determined the shear strength, an attempt was made to measure the distribution of the horizontal displacement at failure and LD strength with the respective applied normal stresses, i.e. 50, 100, 200, 295, and 400 kPa. This relationship is vital, particularly in the design of the landfill as geosynthetics are mostly expected to slip over the adjacent material once they experience the shear stress from the stored waste (Qian 2008; Cilliers 2018b; James 2018a). The horizontal displacement at peak strengths represented the maximum displacement the interface can reach before the decrease in shear stress of the geosynthetic is experienced to reach the LD strength. According to the ASTM D5321 standards, the horizontal displacement at which the

peak and LD strength occur may differ respectively with the specimen gripping system used. In Figure 4-1 the displacement at which the peak and LD occurred were measured and presented in Table 4-1. In all interface tests conducted, the peak and LD strengths were achieved after a certain amount of displacement had taken place. The table showed that the higher the applied normal stresses, the larger was the required displacement to fully mobilize the peak strength. However, the horizontal displacements corresponding to the maximum shear strengths indicated that the NP mobilized the peak strengths at a lower displacement as compared to the SP. The NP peak strengths occurred at a horizontal displacement of approximately 1.1 mm lower than that for the SP as seen in Table 4-1. According to the literature (i.e. Fox et al. 2004), larger displacement value shows that slippage or tension was experienced during shearing. This suggested that the SP interfered with the experiment. Nonetheless, both gripping systems mobilized the peak strength at the horizontal displacement of less than 15 mm which was consistent with the findings of many researchers, including Gilbert & Byrne (1996); Buthelezi (2017); Fox & Ross (2011).

Equally, the NP approached the LD strengths at a much similar displacement as the SP. The LD displacement from the NP was approximately less or equal to 2 mm higher than that of the SP. This suggested that the gripping system had also an effect on LD - displacement.

Table 4-1: Summary of the shear stress versus horizontal displacement results

Normal Stress (kPa)	Gripping Surface	Peak strength (kPa)	Horizontal displacement (mm)	LD strength (kPa)	Horizontal displacement (mm)
50	NP	29	3.3	21.2	70
50	SP	29	4.3	20.6	69
100	NP	63	4.7	39.5	70
100	SP	63	6.5	39.2	70
200	NP	127	5.8	74.9	70
200	SP	116	6.9	61.6	69
295	NP	191	7.6	103	70
295	SP	161	7.5	81.4	69
400	NP	225	6.4	116	68
400	SP	207	8.1	108	66

It was also observed in Figure 4-1 that prior to LD strength, the curves displayed what can be described as a stick-slip behaviour. This behaviour is shown in Figure 4-2.

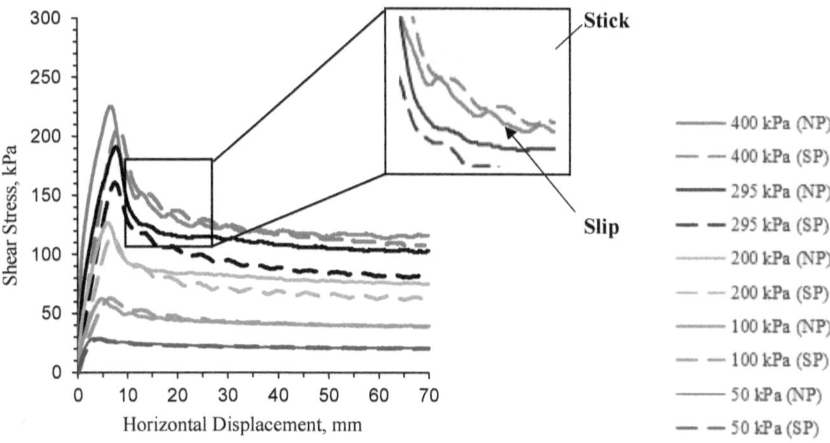

Figure 4-2: Illustration of the stick-slip behaviour in the shear responses

Observations in Figure 4-2 show that at lower stress, 50 and 100 kPa, the curves displaced a post peak steady-state behaviour regardless of the gripping system utilized. At high stress, however, the stick-slip behaviour was noticed with increasing normal stress. Previous researchers, including Soleimanian et al. (2016) and Allen & Fox (2007) have observed this stick-slip behaviour and attributed it to the inadequacy of gripping system i.e. slippage between the test specimens and gripping surfaces. Kalumba (2018), however, related this behaviour to an interlock-slip phenomenon, arising from a repeated process of build-up and subsequent collapse of resistance stresses between the tested interfaces due to the increase in the confining pressure. Comparing the NP shear responses to the SP, the stick-slip behaviour at 200 and 295 kPa, was clearer with the SP than with the NP. This implied that the gripping system had an influence also on this behaviour. At 400 kPa, on the other hand, the shear strain pattern was much similar irrespective of the gripping system, as observed in Figure 4-2.

4.2.3   Geomembrane/Geosynthetic Clay Liner (GGMB/GCL) Interface Test

The envirofix x800, geosynthetic clay liner (GCL) with a thickness range of 2 to 2.7 mm in its un-hydrated state, was used for the tests. The GCL was a reinforced composite product made up of; from top to bottom, non-woven geotextile cover, sodium bentonite layer and woven geotextile

carrier layer as explained in section 3.2. The interface considered in the following test results was the non-woven GCL surface against the GMB side with 1.81 mm asperity heights.

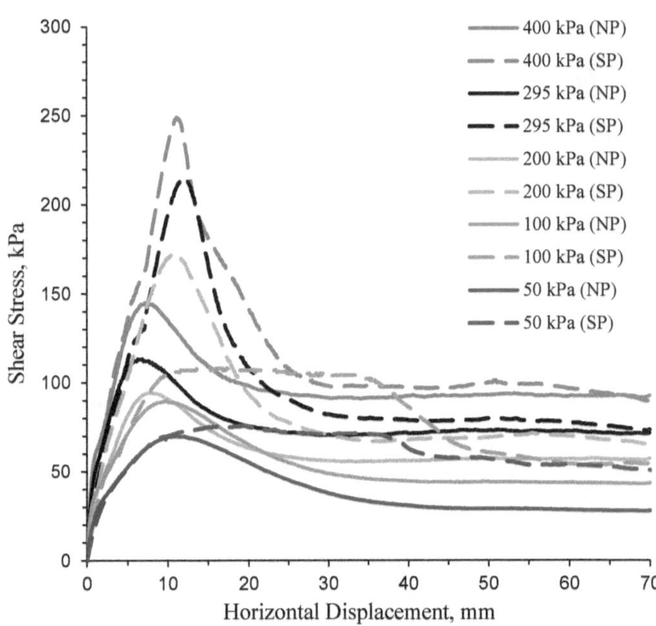

Figure 4-3: Shear stress versus horizontal displacement relationships for GMB/GCL interface

The shear stress versus horizontal displacement relationships for the conducted GTX-A/GCL interface tests are presented in Figure 4-3. The test samples in this experiment were all fully submerged in water for 24 hours before shear was initiated to simulate the least favourable condition of hydration in the landfill, i.e. during rainy season conditions or effluents from stored waste, as explained in section 3.4.4. This hydration was performed under a normal stress of 17 kPa to represent the assumed initial load before waste placement in a landfill. Also, a recommended constant shear rate of 0.1 mm/min was used throughout the GTX-A/GCL experiments to ensure that no excess pore pressures were present on the interface during shear (ASTM D6243 2018).

As in Figure 4-1, both gripping systems in all GMB/GCL interface experiments led to a complete failure of the tested samples based on the observation of shear strength responses and visual

inspection of the failed test samples. The general trend of increasing shear stress with further development of horizontal displacement to reach the peak strengths was identified for any normal stress considered (see Fig. 4-3). Once the peak strengths were achieved, a gradual reduction in shear stresses was noticed with continuous progress in horizontal displacement to approach the LD strengths. At a horizontal displacement of 70 mm, all the tests ended automatically and the measured plots were all non-linear, as shown in Figure 4-3.

The observed shear stress responses exhibited a 'smooth' pre-peak strength in all the normal stresses considered regardless of the gripping method (see Fig. 4-3). However, at normal stresses of 100, 295 and 400 kPa, the curves displayed two distinct pre-peak stress that can be described as a *'skew'* behaviour at different horizontal displacement, depending on the gripping system utilized. This skew behaviour was noticed at displacement between 5 mm and 7 mm for the NP and 0.5 - 2 mm for the SP. This behaviour can be attributed to the interaction between the test samples and the gripping surface. It was anticipated that as the normal stress increased, the gripping systems fully engaged with the tested specimens, thus, developing a strong interlocking bond (Soleimanian et al. 2016). Accordingly, a sharp pre-peak strength was observed after the skew behaviour (see Fig. 4-3). Moreover, at higher applied normal stress, greater interlocking and 'punching' effect of the gripping surface into the test sample fabric i.e. GCL results in an increased resistance to shear at the interface (Allen & Fox 2007).

Beyond the pre-peak strengths, the maximum shear stresses occurred in all the normal stresses considered. The peak strengths were mobilized at a certain amount of horizontal displacement, depending on the used gripping method. While both gripping systems reached the peak strength, the NP measured strength failures were lower than that of the SP in all GMB/GCL experiments conducted. The difference in the peak strengths between the two gripping methods became bigger with increasing normal stresses (see Fig. 4-3). This observation can be related to the difference in gripping systems used in the interface shear test (Fox et al. 2004; Allen & Fox 2007).

It was, however, important to note that at lower stresses, the SP developed a 'steady-state' peak strengths over a range of horizontal displacement before the reduction in shear strength (see Fig. 4-3). This behaviour was again obtained when these tests were repeated at the same applied normal stresses. This steady-state peak strengths behaviour was similar to what Fox et al. (2004) and Rouncivell (2005) presented in their study. The authors suggested that the broader peaks

were due to the test specimens failing in tension which could be related to slippage of the test samples caused by inadequate specimen gripping during shearing (Soleimanian et al. 2016). This implied that the SP gripping surface, consisting of irregular, angular shape particles was only in contact with the test specimens, but did not mobilize sufficient shear resistance between the test sample and the gripping system, resulting in not effectively transferring the applied normal stress to the interface within the tested specimens (Allen & Fox 2007; Stripp 2018).

In Table 4-2 the horizontal displacement required to fully mobilise the peak strengths at all the normal stresses considered, for the two gripping methods were measured and presented. The table showed that as the normal stress increased, the peak-horizontal displacement measured with SP decreased for any particularly normal stress. This horizontal displacement tread was related to sufficient mobilization of the resistance against slippage during shearing. This is because a higher normal stress on the plane of shear failure leads to an increase in interface shear resistance between the tested sample (Hegde & Roy 2018). Accordingly, the peak strengths were mobilised at short displacements. With the NP, however, there was no clear pattern of the horizontal development with increasing normal stresses. The measured NP peak displacement decreased with increase in normal stress, particularly for 50, 100, 200 and 295 kPa, yet at 400 kPa, the horizontal displacement increased from the 6 mm at 295 kPa to 7 mm at 400 kPa for the peak strength to be achieved.

It was also observed in Figure 4-3 that after the peak strengths were mobilized, a decrease in shear stresses (strain softening) was experienced before reaching the LD strength. This strain softening development increased with increasing normal stress when both gripping systems were utilized. It was noticed, however, that the SP showed the most significant drop from the peak to LD strength values (see Fig. 4-3). The strain softening was estimated to range between 37 and 59% for NP and 33 – 66% for SP, in all interface tests conducted.

With the LD strengths attained in each experiment, the effects of the gripping systems on the LD strengths was presented and analysed. It was observed that the NP mobilized a lower LD strength values as compared to that of the SP, for any applied normal stress, as seen in Table 4-2. Although, at a horizontal displacement of about 68mm the shear strength was about the same (crossed curves) for both NP and SP at a normal stress of 400 kPa.

In terms of the horizontal displacement at LD strengths, both gripping methods approached the LD strengths at about 70 mm, as presented in Table 4-2.

Table 4-2: Summary of the shear stress-horizontal displacement results

Normal Stress (kPa)	Gripping Surface	Peak strength (kPa)	Horizontal displacement (mm)	LD strength (kPa)	Horizontal displacement (mm)
50	NP	69.9	11	27.8	69
	SP	75.8	19	51	70
100	NP	89.5	10	44	70
	SP	108	14	54.8	70
200	NP	94.3	8.0	57.2	70
	SP	172	10.4	66	70
295	NP	113	6	71.2	69
	SP	215	12	73	70
400	NP	145	7.0	92	70
	SP	249	11	89.8	69

4.2.4   Geosynthetic Clay Liner/Geotextile-filter (GCL/GTX-B) Interface

A non-woven geotextile-filter (GTX-B), manufactured from virgin polypropylene fibres with added UV stabiliser was used for the tests. This geotextile, made from needle-punched and thermally bonded fibres had a thickness of 0.7 mm. During all the experiments, one respective GTX-B surface was tested against the woven side of envirofix x800 GCL. The resultant shear stress versus horizontal displacement is shown in Figure 4-4.

The investigated shear stress versus horizontal displacement relationships for GCL/GTX-B interface tests was presented in Figure 4-4. Similar to the GMB/GCL interface tests, the GCL and GTX-B test specimen were hydrated for 24 and sheared at 0.1 mm/min to a horizontal displacement of 70 mm.

It is evident from Figure 4-4 that at any particular horizontal displacement, the rate of shear stress development to the failure point for any applied normal stress was not affected by the method of gripping as the shear strengths obtained was much similar for both gripping systems. At maximum shear strengths, however, the deviations in peak strengths between the two gripping systems were observed through the GCL/GTX-B interface shear strength responses. These differences in peak strengths were clearer with higher applied normal stress (seen Fig. 4-4).

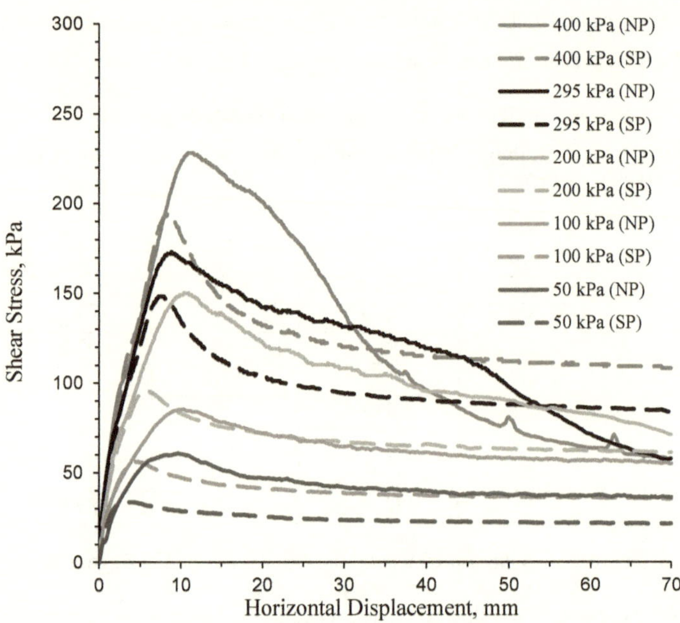

Figure 4-4: Shear stress versus horizontal displacement relationships for GCL/GTX-B interface

Observations in Figure 4-4 show that for any applied normal stress, the peak strength achieved by the NP was higher as compared to that of SP. In Table 4-2 peak strength values obtained from Figure 4-4 are presented. The table shows that the peak strengths mobilised by NP were about 25 - 54 kPa higher than that of the SP, depending on the magnitude of the applied normal stress. According to Fox & Kim (2008), the variation in the mobilized shear strength in geosynthetics interface testing can be mostly related to the effects of the gripping system used during the experiment. This suggested that the method of gripping used during the GCL/GTX-B interface tests exerted an apparent influence on the shear strength mobilized.

The shear stress versus horizontal displacement plots in Figure 4-4 further displayed the strain softening behaviour. The interface shear strength decreased gradually with a continuous development in horizontal displacement in all applied normal stresses. In fact, as observed in Figure 4-4, the strain softening behaviour was greater with higher applied normal stresses. However, two distinct strain softening behaviour was seen with the NP. At higher normal

stresses, the shear strength curves displayed a prolonged strain softening development as compared to a gentle strength reduction at lower normal stresses (see Fig. 4-4). Conversely, the SP displayed consistent strain softening behaviour that increased with increasing normal stresses, as shown in Figure 4-4.

The LD strengths were also obtained for the GCL/GTX-B interface test under similar applied normal stresses and presented in Table 4-3. The LD strength increased with increasing normal stresses irrespective of the gripping methods utilized (see Table 4-3). It is, however, essential to note that the deviation in LD strengths was observed as the normal stress increased. The NP mobilized higher LD strengths until 200 kPa, beyond which the SP had a large LD strength value at 295 kPa and 400 kPa.

According to reviewed literature, variation i.e. high or low in the mobilized shear strength is mainly caused by the difference in the gripping systems used during the experiments. Fox et al. (2004) and Allen & Fox (2007) reported that the reduction of the measured shear strength is mostly caused by slippage of the test specimen over the gripping surface, which leads to the applied normal stress not entirely been transferred within the interface of the tested samples. Higher shear strengths, on the other hand, can be attributed to measuring the yielding strength of the tested material due to the interference of the gripping surface i.e. punching and pulling/ranking of the sample (Kalumba 2018; Hardie 2018a; James 2018b). Reviewing the failed test specimens, visual examination revealed that the NP teeth punctured through the bottom thin grey GTX-B test specimens to 'bite' into the woven surface of the GCL test samples. This caused the surface raking/tearing of the woven surface fabric of the GCL, especially at high applied normal stress (see Fig. 4-5). Based on this, it was suggested that the surface raking/tearing of the GCL resulted in measuring the product 'yield strength' or nail-teeth/GCL interface instead of the GCL/GTX-B interface shear strength (James 2018b; Hardie 2018b). This was because the nail teeth interfered with the test by being in direct contact with the GCL test specimen (as seen from visual inspection).

For the purpose of illustration, Figure 4-5 shows representative test specimens that experienced the surface raking/tearing behaviour. Similar behaviour displayed on a sample tested at 200 kPa was also noticed on specimens tested at normal stresses of 295 kPa and 400 kPa.

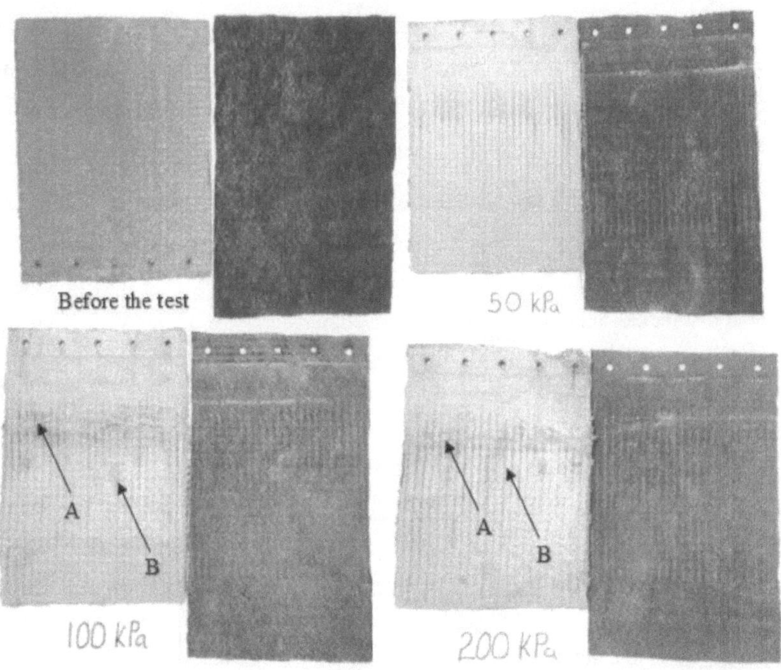

Figure 4-5: Photos of test specimens after the test; A = tearing and B = raking.

In Table 4-3 the horizontal displacement required to mobilize the peak strengths were measured and presented with the corresponding peak strengths. The results show that the displacement required to fully mobilise the failure stress decreased with increasing normal stresses, particularly for the peak strengths obtained using the SP. Contrary to SP, as detailed in Table 4-3, there was no pattern observed in the development of the peak displacement with the increase in normal stress when the NP was utilized. Also, in Table 4-3 is the horizontal displacement at which the LD strength occurred. The NP gripping systems reached the LD strengths at about 2 mm higher than the SP. This suggested that the method of gripping system had an influence also on the mobilized peak and LD horizontal displacements. The shear stress treads behaviours displayed in Figure 4-4 were also obtained from repeatability tests (see Fig. 3-15).

Table 4-3 show the summarized shear strength values obtained at their respective horizontal displacement and normal stresses.

Table 4-3: Summary of the shear stress-horizontal displacement results for GCL/GTX-B interface.

Normal Stress (kPa)	Gripping Surface	Peak strength (kPa)	Horizontal displacement (mm)	LD strength (kPa)	Horizontal displacement (mm)
50	NP	60.8	9.6	36.1	70
	SP	33.9	3.3	21.5	68
100	NP	85.5	10.2	54.8	70
	SP	56.3	4.4	35.3	69
200	NP	150	10.3	71.1	70
	SP	95.8	5.80	61.2	69
295	NP	173	8.80	57.3	70
	SP	148	7.20	83.9	69
400	NP	228	10.8	57.9	70
	SP	194	8.30	108	69

4.2.5 Multi-interface (GTX-A/GMB/GCL/GTX-B) Interface

In a multi-interface test, interfaces are tested simultaneously and allow failure to occur along the weakest interface as anticipated in the field. The peak and LD design strength are determined directly from a single multi-interface failure envelope instead of developing combination peak and LD failure envelopes from the results of several single interface tests, (Stark & Choi 2004). In this study, all the single interface considered were tested at the same time to form a multi-interface test (GTX-A/GMB/GCL/GTX-B). The resultant shear stress versus horizontal displacement is shown in Figure 4-6.

The investigated shear stress versus horizontal displacement behaviour for the GTX-A/GMB/GCL/GTX-B interface test is presented in Figure 4-6. The experiments were conducted at a constant shear rate of 0.1 mm/min to a horizontal displacement of 70 mm after test samples were hydrated for 24 hours under a normal load of 17 kPa.

It was observed, in Figure 4-6 that the measured shear stress responses for the GTX-A/GMB/GCL/GTX-B interface tests were all non-linear for any particularly normal stress applied. The shear strength responses increased with increasing normal stresses to reach the failure points. Once the peak strengths were mobilised, the shear stress in each interface test investigated decreased smoothly to a stable, LD strength under further development in horizontal displacement.

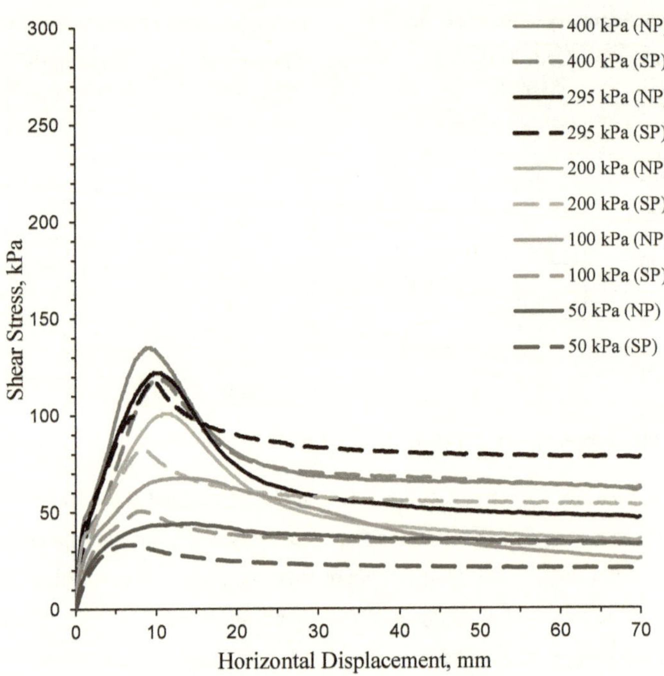

Figure 4-6: Shear stress versus horizontal displacement relationships for multiple interface test

The graphs in Figure 4-6 demonstrated that prior to failure of the tested samples, the rate of development of the horizontal displacement with respect to mobilised shear stress by the NP was lower than that of the SP, for any given normal stress. The shear strength responses achieved with the NP displayed a rapid pre-peak mobilization of shear stresses until reaching the peak strengths. This observed pattern increased with increasing normal stress, however, at normal stresses of 200 and 295 kPa, the development of the shear stress with respect to horizontal displacement was much similar regardless of the gripping method used until reaching the displacement of about 5 mm (see Fig. 4-6).

After pre-peak strengths, the shear stress plots reached the maximum shear strengths at a certain horizontal displacement, depending on the applied normal stress and the gripping system. In Table 4-4, these achieved peak strengths, read-off from Figure 4-6 were presented. It was observed, as details in Table 4-4, that the peak strengths measured using both gripping systems, increased with increasing normal stresses. The NP, however, achieved higher maximum shear strengths in each interface test performed as compared to SP. This difference in the peak strengths between the two gripping methods ranged from 4% to approximately 8%, with the highest being observed in a test conducted at a normal stress of 200 kPa (see Table 4-4). Equally, the lowest peak strength deviation (4%) was at 295 kPa as shown in Table 4-4. These results were much similar to the repeated tests, thus, demonstrated that the method of gripping had an effect on the shear strengths responses even when tested using the multi-interface tests.

The horizontal displacements corresponding to the peak strengths of each shear stress curve in Figure 4-6 were also determined and presented in Table 4-4. The results revealed that the failure stress occurred after developing a certain amount of horizontal displacement, depending on the gripping system. Particularly, as observed in Table 4-4, the required horizontal displacement at any given normal stress to attain the peak strength when measured with the NP, decreased with increasing normal stress. This suggested that the higher the applied normal stress, the larger the distributed confining pressure between the NP and the test samples, thus, resulting in a strong interlocking bond (Allen & Fox 2007; Fox & Kim 2008). Conversely, Table 4-4 displayed that the increase in normal stress resulted in increasing peak horizontal displacement when the SP was used. This may imply that the gripping engagement provided by the SP was not sufficient to transfer the confining pressure to the tested surface of the specimen (Stripp 2018). Moreover, larger displacement values indicate inadequate gripping system during the shear test (Fox et al. 2004). Overall, the peak strengths were generally mobilized at the horizontal displacement between 8.8 mm and 14.4 mm for NP and about 6 – 10 mm for the SP, as shown in Table 4-4.

It was also observed in Figure 4-6 that following the failure points was a drop-off (strain softening) in shear stresses. This reduction in shear stresses was approximately identical in both gripping systems at any applied normal stress (see Fig. 4-6). It was noticed that the strain softening became clearer with increasing peak strengths of the shear responses (see Fig. 4-6). It is, however, essential to note that the rate of reduction in shear stresses was distinct between the two gripping systems. The NP displayed a quick stress softening in all normal stresses except at lower stresses (see Fig. 4-6). At 50 and 100 kPa, the NP had a gradual decrease in shear strength, especially at 100 kPa. On the contrary, at any given normal stress, a gentle stress softening was observed with the SP as shown in Figure 4-6.

In all the five normal stresses, the LD strengths were achieved with further development in horizontal displacement. Similar to the measured peak strengths, the gripping methods also exerted an apparent influence on the LD strengths as observed in Figure 4-6. In Table 4-4, the LD strength values obtained from Figure 4-6 were presented. Compared to SP, a lower LD strength was obtained for the NP at any particular normal stress except for 50 kPa and 295 kPa were NP had a higher and equal LD stress, respectively. The peak strengths obtained from the NP, reduced by 54–65% to reach the LD strengths within the horizontal displacement of approximately 69 mm. Also, as observed in Table 4-4, the NP had mobilized LD strengths of between 0 - 39% lower than the SP, depending on the applied normal stress. The LD strengths occurred within the horizontal displacement of about 68 - 69 mm for the SP.

Table 4-4: Summary of the shear stress - horizontal displacement results for multiple interface test

Normal Stress (kPa)	Gripping Surface	Peak strength (kPa)	Horizontal displacement (mm)	LD strength (kPa)	Horizontal displacement (mm)
50	NP	44.4	14.4	20	70
	SP	33.4	6.1	20.9	68
100	NP	67.7	12.8	25.7	69
	SP	50.7	8.0	32.9	69
200	NP	101	10.8	35	69
	SP	82.6	8.4	53.5	69
295	NP	122	9.9	47	69
	SP	118	9.7	77.5	69
400	NP	135	8.8	62.2	69
	SP	119	10.0	61.9	68

4.2.6 Single versus Multi-interface Tests

The shear stress responses for the multi-interface tests presented in Figure 4-6 contained all of the interfaces which were considered in the single-interface test programme in this study. Hence, the shear stress responses between the single and multi-interface tests can be directly compared.

For all the interface test considered, the shear strength graphs show that before achieving the peak strength, all the experiments demonstrated a pre-peak strength with increasing horizontal displacement regardless of the method of testing. Beyond the peak strength, the strain softening behaviour was observed in both the single and multi-interface tests to reach a steady-state strength. This LD strength was evident in all experiments conducted, as observed in Figure 4-1 to Figure 4-6.

In Table 4-5, a summary of the peak and LD strength obtained from the both single and multi-interface tested was presented. According to Stark et al. (2011), the use of single interface shear tests can lead to an overestimation of the shear resistance of some geosynthetic interfaces. Comparing shear strengths in Table 4-5, higher peak and LD strength values were obtained for interfaces conducted using single interface testing than for multi-interface tests except for tests conducted at a normal stress of 50 kPa were the GTX-A/GMB interface had lowest shear values. This variation in shear strength behaviour is consistent with the finding of Stark et al. (2011). Therefore, suggesting that the interface shear strengths of geosynthetics considered in this study can be influenced by both the gripping systems used in the experiment and the combination method of testing i.e. single or multi-interface.

The reduction in the peak strength for any normal stress considered was also more evident in the single interface tests than in multi-interface, as seen by comparing the single and multi-interface plots.

Table 4-5: Summary of the peak and LD strength obtained from single and multi-interface tests

| Normal Stress | Gripping Surface | Single Interface | | | | | | Multi-interface | |
| | | GTX-A/GMB | | GMB/GCL | | GCL/GTX-B | | | |
		Peak	LD	Peak	LD	Peak	LD	Peak	LD
50	NP	29	21.2	69.9	27.8	60.8	36.1	44.4	20
	SP	29	20.6	75.8	51	33.9	21.5	33.4	20.9
100	NP	63	39.5	89.5	44	85.5	54.8	67.7	25.7
	SP	63	39.2	108	54.8	56.3	35.3	50.7	32.9
200	NP	127	74.9	94.3	57.2	150	71.1	101	35
	SP	116	61.6	172	66	95.8	61.2	82.6	53.5
295	NP	191	103	113	71.2	173	57.3	122	47
	SP	161	81.4	215	73	148	83.9	118	77.5
400	NP	225	116	145	92	228	57.9	135	62.2
	SP	207	108	249	89.8	194	108	119	61.9

4.3 Shear Stress versus Normal Stress

The ASTM D5321/D6243 standards state that a best-fit straight line governed by the Mohr-Coulomb failure envelope can be developed if more than three test points are conducted at

different normal stresses in geosynthetics interface shear tests. In Figure 4-7, the shear strength at peak and LD, read-off from Table 4-5 with the corresponding applied normal stress are connected with a best-fit straight line. The inclination of this line to the horizontal axis corresponded to the interface friction angle of shearing resistance of the tested geosynthetics. The vertical intercept, on the other hand, represented the adhesion.

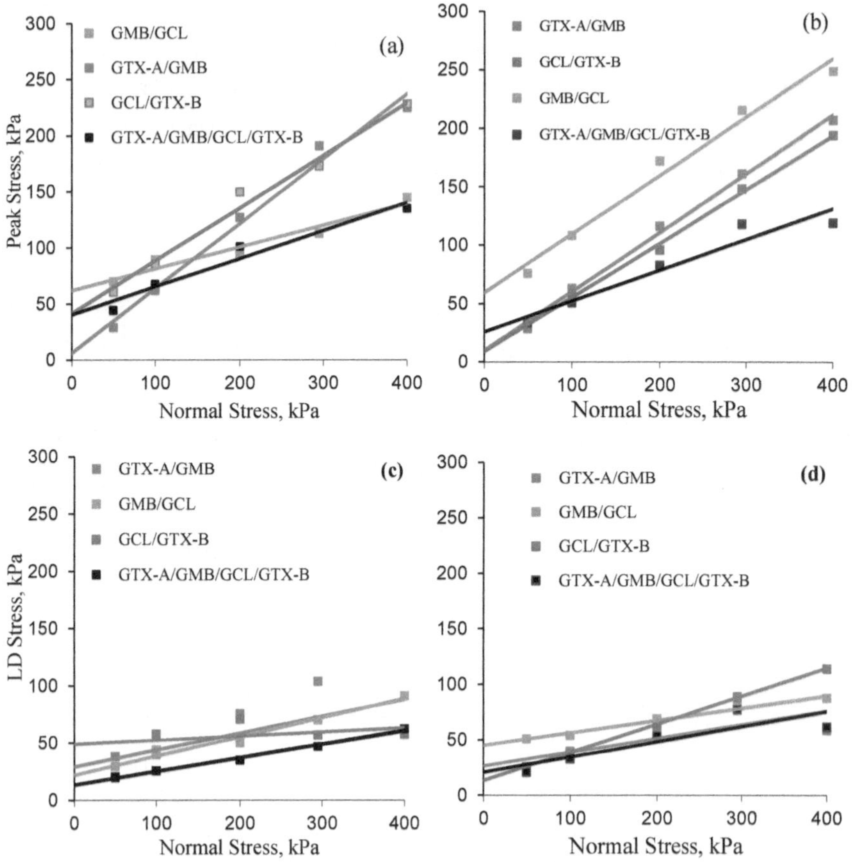

Figure 4-7: Peak stress versus applied normal stress (a) NP and (b) SP (c) NP and (d) SP

In Table 4-6 to Table 4-9, the respective interface friction angles and adhesions obtained from Figure 4-7 are presented, followed by a comparative analysis, as well as a discussion summary.

It is evident, as observed in Table 4-6 to 4-9, that all failure envelope displayed a shear strength intercept (adhesion) regardless of the gripping system used. It was anticipated that the observed adhesion was due to the tested specimens providing some shear strengths even at zero shearing normal stress, as illustrated in Figure 4-7.

4.3.1 GTX-A/GMB Interface

The peak and LD failure envelopes for GTX-A/GMB interface in terms of strength parameters were determined from Figure 4-7 and summarized in Table 4-6. The respective results were calculated by analysing the best-fit straight line governed by the Mohr-Coulomb failure envelope equation.

Table 4-6: Summary of the GTX-A/GMB Interface frictional angle and adhesion

Interface	Gripping system	Friction Angle (°)		Adhesion (kPa)	
		Peak	LD	Peak	LD
GTX-A/GMB	NP	30.0	8.35	6.2	29.3
	SP	26.8	7.08	9.6	26.6

It is evident from Table 4-6 that a difference in friction angle exists between the two gripping systems employed during the interface shear test. NP exhibited a higher peak and LD friction angle as compared to the SP. The peak and LD interface friction angles obtained were 30° and 8.35° for the NP and 26.8° and 7.08° for the SP, respectively. This represented a percentage difference of approximately 11 % and 15 % in peak and LD friction angles for the two types of gripping systems investigated, respectively. This implied that the use of different gripping systems had an influence on the determined interface shear strength characteristics of the geosynthetics considered in the study.

In Table 4-6, the adhesion values of the tested samples were also presented. It was noticed that NP yielded an adhesion value of 6.2 kPa, which is 43% lower than the value obtained using the SP (9.6 kPa). On the contrary, NP mobilized a higher adhesion value at LD strength as compared to that of the SP, (see Table 4-6).

Therefore, based on these results, it was safe to conclude that the use of different gripping systems in geosynthetic/geosynthetic interface shear tests can cause dissimilarities in the measured interface shear strength characteristics.

4.3.2 GMB/GCL Interface

In Table 4-7, a summary of the GTX-A/GMB interface shear parameters determined from Figure 4-7 is presented. Again, the interface friction angles and adhesions were measured using the linear Mohr-Coulomb shear strength envelope.

Table 4-7: Summary of the GTX-A/GMB Interface frictional angle and adhesion

Interface	Gripping system	Friction Angle (°)		Adhesion (kPa)	
		Peak	LD	Peak	LD
GMB/GCL	NP	11.0	9.49	61.9	22.1
	SP	26.6	6.33	59.2	45.1

The results, as detailed in Table 4-7, show that the measured peak and LD interface friction angles with the SP were higher than that of the NP. The SP achieved a peak friction angle of 26.6° which was more than double to what was measured with NP, 11°. Also, the SP had a bigger LD friction angle of 9.49° as compared to the 6.33° achieved with the NP. The percentage difference in both the peak and LD interface friction angle for the SP and NP was 59% and 33%, respectively.

The obtained adhesion values in the conducted interface tests were also included in Table 4-7. Comparison of the presented results shows that the adhesion values at peak, from the NP, was larger than that of the SP. The NP and SP, respectively, had the peak adhesion components of 61.9 kPa and 59.2 kPa, which is approximately 4% in difference. Furthermore, the LD adhesion value of 22.1 kPa was measured with the NP while 45.1 kPa was obtained with the SP.

It should be noted, however, that the observed results from SP should be treated with caution since the test samples experienced some slippage during shearing for the experiments performed at lower stresses (50 and 100 kPa), as detailed in section 4.2.2. This could have interfaced with the normal stress distribution onto the test specimen, hence, forcing shear not to take place on the desired interface (Fox et al. 2004; ASTM D6243 2018; ASTM D5321 2017). This mismatch of results, however, shows the importance of using the appropriate gripping systems in geosynthetics interface shear testing.

4.3.3   GCL/GTX-B Interface

The peak and LD failure envelope derived from the direct shear tests on the GCL/GTX-B interface was shown in Figure 4-7. Similar to the peak shear strength characterises, the LD interface friction angles and adhesions were determined as seen in Table 4-8.

Table 4-8: Summary of the GTX-A/GMB Interface frictional angle and adhesion

Interface	Gripping system	Friction Angle (°)		Adhesion (kPa)	
		Peak	LD	Peak	LD
GCL/GTX-B	NP	25.2	2.1	41.3	48.8
	SP	24.7	14.2	9.25	13.4

In Table 4-8, comparable peak friction angles were observed between the two gripping systems employed. The NP achieved a higher peak friction angle than the SP. The respective peak friction angles from the NP and SP were 25.2° and 24.7°, which is approximately 2 %, percentage difference. Unlike the friction angles obtained at the peak, the LD friction angle for the GTX-A/GMB interface obtained using the NP were found to be significantly smaller as compared to the SP. The NP yielded an LD friction angle of 2.1°, which is 85% lower than the 14.2° mobilized by the SP.

The adhesion results were also presented in Table 4-8 and based on the observation, an apparent difference between the NP and the SP was noticed. The NP yielded large adhesion values both at peak and LD strength than those from the SP (see in Table 4-8). The peak and LD adhesions from NP are 32 and 35 kPa higher than those of the SP, respectively.

Again, the results measured from NP should be treated with caution, particularly tests conducted at normal stresses of 200, 295 and 400 kPa, as the test specimens experienced distortion during shearing as explained in section 4.2.3. It was, therefore, anticipated that the difference in the measured peak and LD friction angles, as well as the adhesions, was related to the influence of the gripping systems used.

4.3.4 GTX-A/GMB/GCL/GTX-B Interface

The interface shear strength parameters for the multi-interface test conducted in this study is shown in Figure 4-7 and respectively summarized in Table 4-9.

Table 4-9: Summary of the GTX-A/GMB Interface frictional angle and adhesion

Interface	Gripping system	Friction Angle (°)		Adhesion (kPa)	
		Peak	LD	Peak	LD

GTX-A/GMB/GCL/GTX-B	NP	14.0	6.77	40.3	13.2
	SP	14.7	7.79	25.9	20.7

From Table 4-9, it was clearly observed that NP mobilised a smaller peak and LD friction angle as compared to SP. The estimated values of the peak and LD friction angles from NP were 14° and 7° which is about 7% and 13% lower than that of the SP.

Similarly, the LD adhesion from the NP was lower than that of the SP. The NP achieved an LD adhesion of 13.2 kPa, while the SP mobilized a value of 20.7 kPa which is about 36% difference. The NP, on the other hand, yielded a higher peak adhesion value as compared to that of the SP. The peak adhesion value obtained by the NP was found to be 40.3 kPa which is about 14 kPa higher than that of the SP. Based on these observations, it can be said that the use of different gripping systems, particularly the NP and SP, in geosynthetic/geosynthetic interface shear tests results in a difference of the measured interface shear strength values.

4.3.5   Summary of the Shear-Normal Stress

Generally, the shear-normal stress graphs of the geosynthetic interface shear strength are as good as the corresponding shear stress versus horizontal displacement relationships. Observing the interface friction angles and adhesion values obtained from shear-normal stress curves in Table 4-6 to Table 4-9, it was evident that there existed a difference in the determined shear strength characteristics. This dissimilarity was attributed to the variation in the gripping systems used during the experiments as all the respective testing condition i.e. procedures, standards, etc. were kept the same (refer to chapter 3). The interaction of the gripping systems with the test specimens has been explained in section 4.2. It was also important to note that the greatest difference in the determined shear strength parameters was observed when the GCL interface was involved, as detailed in Table 4-6 to 4-9, thus, suggesting that the GCL was more susceptible to the choice of the gripping systems than other geosynthetics used i.e. GTX-A and GMB.

In Table 4-10, a summary of the appropriate gripping systems to use in each interface, depending on which parameter is of interest i.e. interface friction angle or adhesion value is suggested. The selection of the respective gripping system was based on which gripping system displayed a higher representative value of the shear strength based on section 4.2 and 4.3 result analysis and discussion.

Table 4-10: Summary of the appropriate gripping system in each interface considered

Interface	Appropriate gripping system	
	Friction angle	**Adhesion value**
GTX-A/GMB	NP	NP
GMB/GCL	NP	NP
GCL/GTX-B	SP	SP
GTX-A/GMB/GCL/GTX-B	SP	SP

4.4 Failure Envelope

The failure envelopes in this study were approximated using a straight fitted line which was analysed using the linear Mohr-Coulomb equation. It was, however, recognised from Figure 4-7 that a nonlinear (i.e. bilinear or curvilinear) model could represent the failure envelope more appropriate. Moreover, Jogi, (2005) highlighted the importance of investigating the nonlinearity nature of the shear - normal stress graphs of geosynthetics, especially when tested over a wide range of normal stresses. This section, therefore, attempted to evaluate the nonlinearity of the shear strengths of the tested geosynthetics.

To achieve this, all best-fit straight lines in Figure 4-7 with linear regression (R^2) of less than 0.98, were represented with nonlinear failure envelope in Figure 4-8. The best-fit straight lines (dashed line) were also included in the graphs for comparison purposes. The R^2, according to Minitab Inc (2018), is a statistical measure of how close the data points are to the fitted regression plot. In this study, the $R^2 \geq 0.98$ was considered to be a well-fitted and best representation of the graph (Minitab Inc 2018; Kalumba 2018).

In this study, as seen in Figure 4-8, only 37.5% of the data points were best-fitted ($0.98 \leq R^2 \leq 1$) using a linear failure envelope and 62.5% wbybest represented with nonlinear (i.e. bilinear and curvilinear) failure envelopes, irrespective of the gripping system employed. Therefore, there was a need to consider nonlinearity failure envelopes to report appropriate normal stress ranges with interface strength parameters obtained using linear equations (Triplett & Fox 2001).

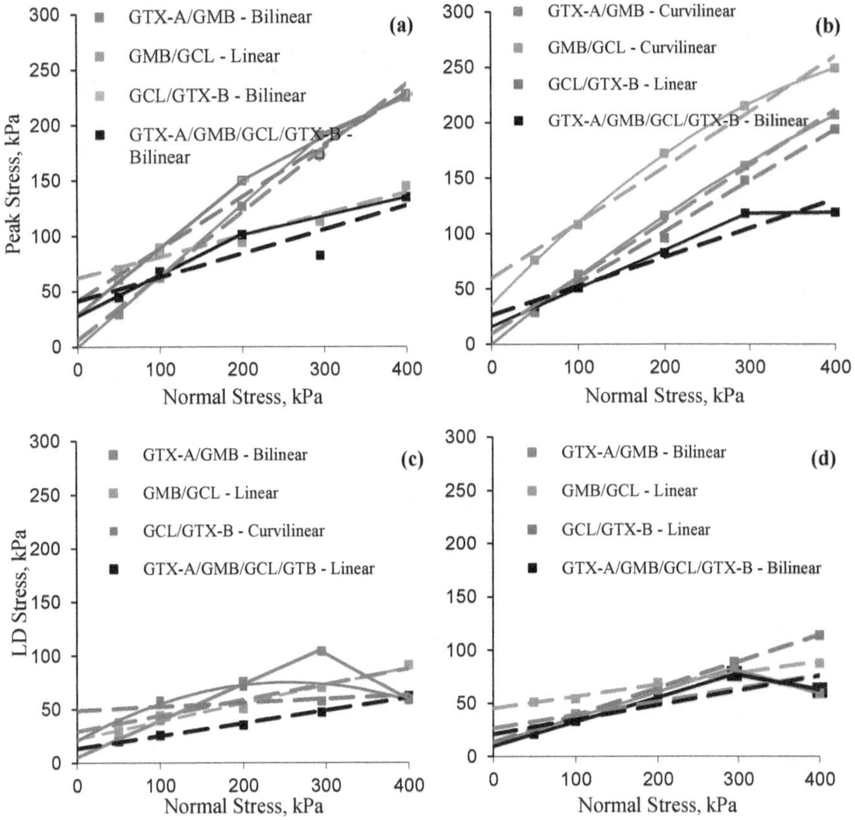

Figure 4-8: (a) Peak vs normal stress – NP (b) Peak vs normal stress - SP (c) LD vs normal stress - NP and (d) LD vs normal stress – SP.

Observation in Figure 4-8 shows that two types of nonlinear curve i.e. bilinear and curvilinear were generated. The selection of which nonlinearity curve to use was based on how close the respective graph is to the data points i.e. a nonlinearity curve with a higher R^2 was preferred to represent the failure envelope.

A bilinear failure envelope, as seen in Figure 4-8 demonstrated that the shear strengths increased with increasing normal stress until a certain normal stress termed, critical stress was reached. Beyond this critical stress, the rate of shear strengths suddenly changed, concave downwards, with the increase in the applied normal stress, resulting in a bilinear curve. As a consequence, an

alteration in the interface friction angle and adhesion after the critical stress was experienced. The critical normal stresses were found to range from about 200 kPa to 295 kPa, depending on the interface shear tested. To determine the shear strength characteristics, equation 3-2 was used, but fitted twice on separate smaller shear stress ranges i.e. 0-295 kPa and 295-400 kPa. Accordingly, two interface friction angles and adhesion values were obtained.

Similar to bilinear curves, curvilinear failure envelopes displayed that the shear strengths increased with increasing applied normal stress, but only up to a certain shear stress point, after which the rate of shear strength with normal stress changed gradually in the concave downwards direction as observed in Figure 4-8. This curvilinear failure envelope implied that the interface friction angle at any point on the failure envelope curve was unique for any applied normal stress, respective of the gripping system utilized. This type of failure envelope was, therefore, modelled by equation 3-3 presented in section 3.7.1.

The nonlinearity of the geosynthetic failure envelope has been observed by other researchers and was attributed to the change in the interaction mechanisms of material tested (Buthelezi 2017; Bacas et al. 2015). Looking at Figure 4-8, however, it was important to note that the gripping system effect also influenced this behaviour. It was evident that based on the experimental data points obtained using the NP, 50% were best fitted using a bilinear failure envelope while 12.5% were fitted as curvilinear. With the SP, however, 37.5% were best fitted as bilinear failure envelope and 25% as curvilinear. Therefore, it can be said that the gripping system has an effect on the shear strength results irrespective the failure envelopes utilized.

In Table 4-11 to 4-12, the summary of the peak and LD strength parameters obtained from Figure 4-8 was presented. The subscripts 'b$_1$' and 'b$_2$', on the interface friction angle terms and 'a$_1$' and 'a$_2$' on the adhesion teams, represented the determined values before and after the critical stress on the bilinear failure envelope, respectively. Also, the constant coefficients obtained through a good fitting ($0.99 \leq R^2 \leq 1$) curvilinear failure envelope technique were presented in Table 4-13. The negative values obtained on the 'a' coefficient indicated that the curve would bend concave downwards if the normal stresses were further to be increased i.e. beyond 400 kPa (Crystal Clear Maths 2013).

Table 4-11: Summary of the peak interface friction angle and adhesion obtained from Figure 4-8

Interface		Friction Angle (°)	Adhesion

	Gripping Surface	ϕ	ϕ_{b1}	ϕ_{b2}	c_a	c_{a1}	c_{a2}
GTX-A/GMB	NP	30.0	32.7	17.9	6.2	0	95.5
	SP	26.8	-	-	9.6	-	-
GMB/GCL	NP	11.0	-	-	61.9	-	-
	SP	26.6	29.8	17.9	59.2	-	-
GCL/GTX-B	NP	25.2	31.0	21.3	41.3	28.6	72
	SP	24.7	-	-	9.25	-	-
GTX-A/GMB/GCL/GTX-B	NP	14.0	20.4	9.65	40.3	27.8	67
	SP	14.7	18.9	0.54	25.9	16.0	115

Table 4-12: Summary of the LD interface friction angle and adhesion obtained from Figure 4-8

Interface	Gripping Surface	Friction Angle (°)			Adhesion		
		ϕ	ϕ_{b1}	ϕ_{b2}	c_a	c_{a1}	c_{a2}
GTX-A/GMB	NP	8.35	18.8	-	29.3	5.25	-
	SP	7.08	13.5	-	26.6	12.3	-
GMB/GCL	NP	9.49	-	-	22.1	-	-
	SP	6.33	-	-	45.1	-	-
GCL/GTX-B	NP	2.1	-	-	48.8	-	-
	SP	14.2	-	-	13.4	-	-
GTX-A/GMB/GCL/GTX-B	NP	6.77	-	-	13.2	-	-
	SP	7.79	12.8	-	20.7	9.43	-

Table 4-13: Coefficient constants obtained from the curvilinear failure envelopes

Interface	Gripping Surface	Peak strength			LD strength		
		a	b	c	a	b	c
GTX-A/GMB	SP	-0.0003	0.6579	-	-	-	-
GMB/GCL	SP	-0.0007	0.8201	35.343	-	-	-
GCL/GTX-B	NP	-	-	-	-0.0008	0.4266	20.617

4.5 Critical Interface

The primary objective of the shear stress versus normal stress graph is mainly to determine the critical/weakest interface of the tested materials i.e. geosynthetics. This is because finding the weakest interface, is key in the stability analysis of a structure with multilayered geosynthetic liner i.e. landfills (Qian & Koerner 2004).

A critical interface, according to Cilliers (2018b), is the surface having the least shear resistance and is likely to cause failure of a structure if its interface friction is exceeded. Therefore, locating this interface in a multilayered geosynthetics liner i.e. landfill is imperative, as it governs the stability design analysis (Cilliers 2018a).

It has been reported, by Stark et al. (2011), that geosynthetic interfaces exhibit stress-dependent shear resistance and as a result, the critical interface varies with the applied normal stress. Hence, determining the weakest interface involves the evaluation of each interface component of a multilayered liner (Cilliers 2018c).

In landfills, the weakest interface has been mainly conducted using single interface tests and little has been reported on a comparison with multi-interface shear tests (Stark et al. 2011). Under this section, a comparison of the peak and LD failure strength envelopes from single and multi-interface direct shear tests presented in Figure 4-8 was considered. These results were particularly preferred as they provided the best fit ($0.98 \leq R^2 \leq 1$) of the failure envelopes, as mentioned in section 4.4.

4.5.1   Peak strength

4.5.1.1   NP Gripping System

Figure 4-8a displayed the strongest and critical/weakest interface shear plan tested. The strongest and critical interface is the interface plotted above and below on the shear strength versus normal stress graph, respectively (Cilliers 2018c). It was evident from Figure 4-8a that the strongest interface changed with the increase in normal stresses, particularly for the single interface tests. The GMB/GCL was the strongest interface at low applied normal stress, i.e. less than 100 kPa. Once the normal pressure exceeded the normal stress of 100 kPa, the strongest interface switched from the GMB/GCL to the GCL/GTX-B. Equally, a similar pattern was observed in the weakest interface shear plan. The GTX-A/GMB and GMB/GCL interface from the single tests exhibited the lowest peak strength and were the critical interface as shown in Figure 4-8a. The GTX-A/GMB displayed the weakest interface for the normal stresses below or approximately 140 kPa whereas the GMB/GCL interface was the lowest for the normal stresses greater or equal to 140 kPa.

Compared to the single interface results in Figure 4-8a it was observed that the multi-interface results exhibited the critical interface for much of the normal stresses considered in the investigation. It was clear from Figure 4-8a that the weakest interface changed from GTX-A/GMB to GTX-A/GMB/GCL/GTX-B at a normal stress greater than 120 kPa with the GTX-A/GMB interface been critical at normal stresses below 120 kPa. The GTX-A/GMB/GCL/GTX-B was critical for normal stresses exceeding 120 kPa, hence, the peak failure envelope corresponded to the GTX-A/GMB/GCL/GTX-B for normal stresses greater than 120 kPa. Therefore, it was necessary to construct a failure combination strength envelope using the two methods of interface testing (single and multi-interface) to represent the critical interface for the normal stresses considered.

4.5.1.2 SP Gripping System

The single interface tests from Figure 4-8b showed that the GMB/GCL interface yielded the highest peak strength envelope in all peak interface test considered. However, as highlighted in section 4.2, the GMB/GCL interface results should be taken with caution as there was interference of the SP with the experiment at low normal stresses i.e. 50 and 100 kPa, hence, could have interfaced with the results obtained. Even so, this showed the importance of using the appropriate gripping system in geosynthetic interface shear testing.

The lowest interface, in Figure 4-8b, displayed a stress-dependent shear resistance with the increase in the applied normal stresses. For instance, the GTX-A/GMB interface was the lowest and critical interface for the normal stress between 0 kPa and 65 kPa whereas, the GCL/GTX-B interface was the weakest interface for normal stresses exceeding 65 kPa.

Comparing the weakest interface failure envelope obtained from the single to the multi-interface test, similar tread as seen in Figure 4-8a was observed in Figure 4-8b. The GTX-A/GMB interface exhibited the lowest and critical interface for the normal stresses below or equal to 65 kPa, while the GTX-A/GMB/GCL/GTX-B was the critical for normal stresses greater or equal 65 kPa.

4.5.2 LD strength

4.5.2.1 NP Gripping System

The LD failure combination strengths determined using the NP gripping system are presented in Figure 4-8c. The single interface test results showed, as observed in Figure 4-8c, that the critical interface varied with normal stress. The GMB/GCL interface exhibited the highest interface for normal stresses below 110 kPa and the GTX-A/GMB interface was the strongest for normal stresses greater than 110 kPa.

The lowest interface interchanged between the GTX-A/GMB and GMB/GCL interfaces, as observed in Figure 4-8c. The GTX-A/GMB was the lowest interface for normal stresses below 110 kPa, hence, the critical interface. The GCL/GTX-B interface, however, was the weakest interface for stresses greater 110 kPa as illustrated in Figure 4-8c.

Relating the single to the multi-interface failure envelops, it was observed that the GTX-A/GMB/GCL/GTX-B interface exhibited the lowest LD strength and was the critical interface for all the normal stresses considered except at 50 kPa and below (see Fig. 4-8c). The GTX-A/GMB interface yielded the weakest interface for normal stresses at 50 kPa and below.

4.5.2.2 SP Gripping System

In Figure 4-8d, the LD failure strength combinations obtained using the SP gripping method are displayed. The strongest interface was found to be a function of normal stress, thus, it changed with increasing normal stresses. For the normal stresses from 0 kPa to 240 kPa, the GMB/GCL interface exhibited the highest interface. Beyond the 240 kPa, the GCL/GTX-B interface displayed the strongest interface as it can be seen in Figure 4-8d. Conversely, the GTX-A/GMB interface test was the weakest single interface measured in all full range of normal stresses considered, as observed in Figure 4-8d. Nonetheless, for the normal stresses between 100 kPa and 200 kPa, the GCL/GTX-B yielded essentially the same LD strength as the GTX-A/GMB interface.

Comparing the weakest interface failure envelope obtained from the single to the multi-interface test response, it was found that the multi-interface test yielded the weakest interface for all normal stresses considered, as observed in Figure 4-8d.

4.5.3 Summary of the critical interface

In Table 4-14, the summary of the critical/weakest interface was presented based on the result analysis in 4.5. The strongest interface was also included for completeness. It is evident from Table 4-14, that the strongest interface was obtained from tests conducted using single interface tests. The weakest interface, on the other hand, was found to vary between the single and multi-interface tests, even though it was more dominant with multi-interface tests. Therefore, based on these findings, it was necessary to construct a failure combination strength envelope using the two methods of interface testing (single and multi-interface) to represent the critical interface for the normal stresses considered.

Table 4-14: Summary of the strongest and critical interface

Interface	Gripping system	Strongest Interface	Weakest Interface
Peak Strengths	NP	Single	Single/Multi-Interface
	SP	Single	Single/Multi-Interface
LD strengths	NP	Single	Single/Multi-Interface
	SP	Single	Multi-Interface

5   PRACTICAL APPLICATIONS

5.1   Introduction

In this chapter, the practical application of the research findings was discussed with the help of a design example of a landfill lining system. The main objective of the chapter was to demonstrate the impact of using the results obtained from one gripping system i.e. nail plate (NP), as opposed to the other, i.e. sandpaper (SP). To achieve this, the shear strength characteristics obtained from shear interface tests i.e. geotextile-cushion (GTX-A)/geomembrane (GMB), GMB/Geosynthetic clay liner (GCL), GCL/Geotextile-filter (GTX-B) and GTX-A/GMB/GCL, in this study were selected. Accordingly, the landfill was assumed to consist of a liner system with multilayered geosynthetic including the geotextiles, geomembranes and GCL.

Locating the critical interface is the most significant aspect of the slope stability analysis of multilayered liner landfills. This involves the evaluation of each interface component of a liner system (Cilliers 2018c). For the purposes of illustration, however, the critical interfaces were assumed to be between geosynthetics only as they are mainly considered to form the critical interface in landfill liners (Visser 2018). Therefore, only results presented in this study were selected in the analysis, nonetheless, the principles used in the design apply equally to other landfill liner interface components (Qian et al. 2003; Qian 2008).

Many researchers including Rouncivell (2005) and Dookhi (2013), have analyzed the liner stability of a landfill using the results determined from a liner Mohr-Coulomb failure criterion. This is consistent with the recommendation from ASTM D5321/D6243 standards as well as from Koerner & Koerner (2007). Moreover, the landfill is assumed to be subjected to lower and higher confining stresses at different stages of construction, filling and post-closure, thus, a linear Mohr-Coulomb failure envelope would be appropriate to represent the failure mode (Buthelezi 2017). In this chapter, therefore, the results obtained from the NP and SP by assuming straight Mohr-Coulomb failure envelopes (refer to section 4.3) were utilized. The respective results were obtained from interface shear experiments conducted at a normal stress range of 50 - 400 kPa. The normal stresses were calculated based on the anticipated field conditions, as explained in section 3.5.

5.2   Design Method

The design of the landfill liner system has evolved with the increase in environmental regulations, siting hearings, and increased public awareness. This has resulted in a modern landfill being constructed of mainly geosynthetic components. Generally, a modern landfill consists of three major engineering elements, basal liner, side slope liner and capping liner, as categorized by the Geosynthetics Interest Group of South Africa (2018). These elements are explained in section 2.3.1 of this study, hence, this chapter focused solely on the application of the results on these elements, particularly the basal liner. This is because the instability of the basal liner is mainly associated with landfill failures (Qian et al. 2003; Cilliers 2018b).

In landfill stability analysis, i.e. basal, slope, etc., the value of the factor of safety (FS), remains the key finding to approximate how close or far the liner is from failure (Hammah et al. 2009). This estimation can be done by many numerical methods, but the traditional limit equilibrium method is the most used (Griffiths & Lane 2001), for that matter in South Africa (Dookhi 2013). This is because the limit equilibrium method is considered to be conservative and provides a straightforward approach when analysing multi-linear shear strength (Baba et al. 2012; Rabie 2014; Rouncivell 2005).

In this book, a formula proposed by Qian et al. (2003) and modified by Qian & Koerner (2004, 2005, 2007) and (Qian 2006, 2008) was used to analyse the basal stability of the considered landfill liner example. According to Qian et al. (2003), failure in a landfill can happen in various modes including rotational failure, translational failure and composite failure. However, the translational mode of failure is the most likely failure type to occur in many modern landfills (Ruan et al. 2013; Qian et al. 2003). In Figure 5-1, a typical example of a translational failure is illustrated and based on this figure, Qian et al. (2003) derived a two-part wedge method for translational failure analysis using the limit equilibrium method. The method estimates a representative value of the FS that ensures that the strength of the waste is not exceeded anywhere within the waste mass (Qian et al. 2003). The principles and other assumptions involving this method are highlighted in section 2.3.1.3 of this study.

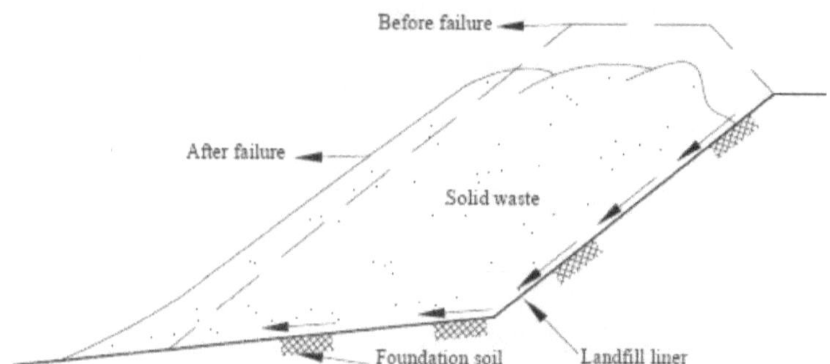

Figure 5-1: Translational waste mass sliding completing along/within a liner system (after Qian et al. 2003).

In order to illustrate the effects of the gripping system on the shear strength results obtained in this study, Figure 5-2 and Table 5-1 presented a proposed 'standard' situation for the analysis based on Qian & Koerner (2004). The unit weight of solid waste, however, was based on the assumed valued used to determine the applied normal stress used in the experiments to ensure that the analysis was in accordance with shear strength properties. Therefore, the design outcome was limited to input parameters considered herein.

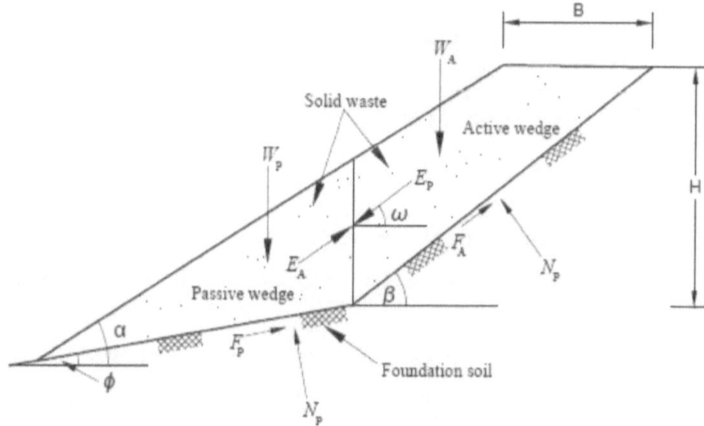

Figure 5-2: Forces acting on two adjacent wedges of a waste mass in a landfill cell (after Qian et al. 2003)

Table 5-1: Adopted control properties for analysis (Qian & Koerner 2004).

Property	Abbreviation	Units	Value
Angle of front slope		°	14.0
Angle of back slope		°	18.4
Height of back slope	H	m	30
Top width of the waste mass		m	20
Unit weight of solid waste*		kN/m^3	9.81
Angle of landfill cell subgrade		°	1.1
Internal friction angle of solid waste	ϕ_{sw}	°	33

*Based on the assumed site condition (Visser 2018).

Simplifying Assumptions

The correctness of a design, according to Ongodia (2017) is limited to the geometry, construction method, theoretical basis of the solution, the interpretation of computed results, support parameters selected and assumptions made. Although Table 5-1 provided the adopted properties of the design, the following additional assumptions were made in the analysis:

- The landfill foundation conditions were deemed adequate for construction.

- The groundwater table was way below the foundation soil layer as well as the leachate level was zero, hence, the influence of pore water pressures was not considered (Qian et al. 2003; Buthelezi 2017).

- The waste mass was to be placed on a multilayered geosynthetic liner system consisting of geotextiles, geomembranes and GCLs, hence, the design was limited by an assumed critical failure surface between these geosynthetic (Visser 2018).

- The landfill liner would be subjected to lower and higher confining pressures at different stages of construction, thus, a linear Mohr-Coulomb failure envelope was chosen for this design (Buthelezi 2017).

- According to Koerner & Koerner (2007), in order to utilize an adhesion value in the design that involves geosynthetics, a clear physical justification has to be made. Therefore, the cohesion of the solid waste and the interface adhesion between two geosynthetic surfaces were omitted due to the uncertainty in these parameter values.

Moreover, ignoring them simplifies the mathematical analysis and will tend toward a conservative design analysis (Qian et al. 2003; Koerner & Koerner 2007; Cilliers 2018b).

- The potential translational failure surface in the liner system would pass through the same interface at both the back slope and base (Qian et al. 2003; Buthelezi 2017).

- All the design parameters remained constant throughout the analysis except the respective interface friction angle of the geosynthetics (Qian et al. 2003).

- The design was considered stable if the $FS \geq 1.5$ when the peak strength values were used. Equally, a $FS \geq 1.0$ presented a stable design if the LD strength values were utilized (Cilliers 2018b).

5.3   Design Output

The design parameters presented in Table 5-1, as well as the friction angles in Table 5-3, were implemented into Equation 2-1 to Equation 2-8 to determine the minimum FS as shown in Table 5-2. For the purpose of illustration, as well as to avoid repetition, only calculation for the GTX-A/GMB interface measured using the NP were detailed in Table 5-2. The design principle, however, applies equally to all the other investigated interface, regardless of the gripping system used. In Table 5-3, a summary of the resultant FS for all the interfaces considered in this study is shown.

Also, in Table 5-3, a percentage difference between the calculated factor of safety (% FS_{PD}) and the recommended FS $\left(FS_{min} \leq 1.5\right)$ was presented. This variation demonstrated how far or close the mobilized FS from the respective gripping system i.e. NP or SP, is from the recommended value.

Table 5-2: Design calculations

Interface	Formula (Qian et al. 2003; Qian & Koerner 2004)	Output	Remarks
GTX-A/GMB	For $B < \dfrac{H}{\tan\beta}$ $C_A = c_a \cdot \dfrac{H}{\sin\beta}$	$C_A = 0$	c_a is assumed to be 0
	$W_A = 0.5 \cdot \gamma_{sw} \cdot \dfrac{H^2}{\tan\beta} - 0.5 \cdot \gamma_{sw} \cdot \left(\dfrac{H}{\tan\beta} - B\right)^2 \cdot \tan\alpha$	$W_A = 7534.7$	
	$C_P = c_p \left[\left(H - \dfrac{H.\tan\alpha}{\tan\beta} + B.\tan\alpha \right) \div (\cos\theta.\tan\alpha - \sin\alpha) \right]$	$C_p = 0$	c_p is assumed to be 0
	$W_p = 0.5 \cdot \gamma_{sw} \cdot \left[\left(\dfrac{H}{\tan\alpha} - \dfrac{H}{\tan\beta} + B \right)^2 \cdot \dfrac{\tan\alpha.\tan\theta}{\tan\alpha - \tan\theta} + \left(\dfrac{H}{\tan\alpha} - \dfrac{H}{\tan\beta} + B \right)^2 \cdot \tan\alpha \right]$	$W_p = 3463.5$	
	$a = W_A.\sin\beta.\cos\theta + W_p.\cos\beta.\sin\theta$	$a = 2441$	
	$b = \left(W_A.\tan\delta_p + W_p\tan\beta\right)\sin\beta.\sin\theta$ $\quad - \left(W_A.\tan\delta_a + W_p.\tan\delta_p\right)\cos\beta.\cos\theta - C_A.\cos\theta - C_p.\cos\beta$	$b = -6208.7$	
	$c = -\left[\begin{array}{l} \left(W_A\cos\beta.\sin\theta + W_p.\sin\beta.\cos\theta\right)\tan\delta_a.\tan\delta_p \\ + C_A.\sin\theta.\tan\delta_p + C_p.\sin\beta.\tan\delta_a \end{array} \right]$	$c = -410.1$	
	$FS_{min} = \dfrac{-b \pm \sqrt{b^2 - 4 \times a \times c}}{2 \times a}$	$FS_{min} = 2.6$	
			$2.6 > 1.5 = FS_{min}$, **ok**

Table 5-3: Variation in Factors of safety for each interface.

Interface	Gripping system	Friction Angle (°)	FS	% FS$_{PD}$*
GTX-A/GMB	NP	30.0	**2.6**	**42**
	SP	26.8	**2.3**	**34**
GMB/GCL	NP	11.0	**0.9**	**-66**
	SP	26.6	**2.3**	**34**
GCL/GTX-B	NP	25.2	**2.1**	**28**
	SP	24.7	**2.0**	**25**
GTX-A/GMB/GCL/GTX-B	NP	14.0	**1.1**	**-36**
	SP	14.7	**1.2**	**-25**

*Percentage difference between the calculated FS and the recommended FS. Negative values show that the value is below the minimum FS, thus, illustrated instability.

The use of peak values at the crest of a slope in landfill stability analysis remains the choice of the designer or the engineer. However, solid waste generally settles considerably during and after the filling operation. According to Spikula (1997), the solid waste stored in the landfill loses approximately 10 - 30% of the initial height as they settle and this can result in the induced shear stress in the liner system i.e. on the side slope, which can lead to shear displacement of the liner system downslope (Qian 2008). As a consequence, shear displacements along specific interfaces in the liner system are induced by shear stresses, resulting in the mobilization of a reduced or residual interface strength (Qian 2008). Additionally, Stark and Poeppel (1994) and Qian et al. (2002) reported that thermal expansion and contraction of the side slope liner system during construction and waste filling may also contribute to the accumulation of shear displacements and the mobilization of a residual interface strength (Qian 2008). Based on these findings, the use of residual interface shear strengths at the slope was suggested for the liner system in the landfill stability analysis of many important landfill projects (Qian et al. 2008). In this study, however, the post-peak shear strength values were considered as LD strength as explained in section 2.3.2.1. Therefore, Table 5-4 shows the analysis results based on the LD values determined in this research. The same calculation procedures as in Table 5-2 but with peak values replaced with LD values, were followed to calculate the FS.

Table 5-4: Variation in Factors of safety for each interface.

Interface	Gripping system	Friction Angle (°)	SF	% FS$_{PD}$*
GTX-A/GMB	NP	8.35	**0.7**	**-30**
	SP	7.08	**0.6**	**-40**

Continuation of Table 5-4

Interface	Gripping system	Friction Angle (°)	SF	% FS$_{PD*}$
GMB/GCL	NP	9.49	0.8	-20
	SP	6.33	0.5	-50
GCL/GTX-B	NP	2.1	0.2	-80
	SP	14.2	1.1	10
GTX-A/GMB/GCL/GTX-B	NP	6.77	0.5	-50
	SP	7.79	0.6	-40

5.4 Discussion of the Design Output

It was observed, as details in Table 5-3 that much of the results (about 63%) obtained from both the NP and SP achieved the recommended FS value (i.e. $FS \geq 1.5$). However, the point of the analysis was to illustrate that the variation in the shear strength parameters determined using different gripping system i.e. NP and SP, on a large direct shear device, can lead to different FS. As a consequence, designers can experience difficulties in their interpretation.

It is, however, important to note that in cases where the FS does not meet the minimum recommended value i.e. the GMB/GCL and GCL/GTX-B interface, the variation in FS can lead to serious complications. For instance, if the GMB/GCL was the critical interface, the designer would use the FS of 2.1 (determined with SP) in the stability analysis while the other would use 0.9 (determined with NP). The latter designer is likely to adjust the design properties i.e. material properties and landfill geometry in order to achieve the recommended FS. This may lead to cost implication (i.e. addition of reinforcement material) and delays of the project (if materials are changed, shear parameters have to be tested in the laboratory). Conversely, if the FS of 2.1 was used and it happens that it was not the actual representative value of the material, it may lead to instability of the landfill structure. This may cause the failure of the landfill and consequently, loss of lives and properties (Koerner & Soong 2000; Stark 2018). Equally, the same can be said about the GCL/GTX-B interface results as well as the results in Table 5-4.

Therefore, obtaining accurate and reliable laboratory generated shear strength parameters (i.e. friction angle and adhesion) of geosynthetics for projects that involve geosynthetic liners i.e. like landfill liner system, is a crucial aspect in preventing failure (Qian 2008).

6 CONCLUSIONS AND RECOMMENDATIONS

6.1 Introduction

In this study, the effects of using different specimen gripping systems in the shear strength at the geosynthetic/geosynthetic interface in a landfill liner application were investigated using the ShearTrac-III, large direct shear device with a square dimension of 305 mm. A comprehensive test program was established to include two commonly used gripping systems i.e. nail plate (NP) and sandpaper (SP), in the respective tests.

The research approach developed in this work enabled the investigation of the shear stress responses, shear strength versus normal stress plots, appropriate failure envelope and the determination of the critical interface in a landfill liner. This analysis provided key findings of the importance of using one gripping system i.e. NP, as opposed to the other i.e. SP in determining the geosynthetic shear strength characteristics at the interface. This chapter, therefore, provided a summary of the conclusions drawn from the findings emerging from this study, followed by areas where further efforts motivated by the research would be required.

6.2 Summary of Conclusions

The conclusions presented herein were derived from this study:

1. To date, there is no standardized gripping system used on a large direct shear device during the geosynthetic/geosynthetic interface shear strength testing. This has led researchers and many laboratories, employing different gripping systems during the experiments. The NP and SP, are one of the commonly used gripping systems and were studied herein. The research results revealed that the geosynthetic interface shear strength characteristics i.e. interface friction angle and adhesion are dependent on the specimen engagement with the gripping system and the applied normal stress. This is because the actual friction angle and adhesion of the considered interface are determined when the intended failure surface has the least shear resistance of all possible sliding surfaces. This is achieved when the tested specimen is appropriately secured to the shearing blocks of the direct shear apparatus, such that the applied normal stress is effectively transferred within the specimen.

2.   In all interface tests conducted, both the NP and SP exhibited typical geosynthetic shear stress responses with nonlinear behaviour, regardless of the applied normal stress. It was, however, noticed that with the increasing normal stress, deviation in the mobilized shear stress between the two gripping systems increased irrespective of the interface tested. However, the exact percentage difference in shear strength parameters obtained when one gripping system is used in relationship to the other was not established as it varied depending on the interface.

3.   The use of a linear Mohr-Coulomb failure envelope to determine the interface shear strength characteristics between two geosynthetic can underestimate or overestimate the shear strength. Therefore, the nonlinearity analysis i.e. bilinear and curvilinear of the interface shear strength failure mode should be highly considered, especially if the experiment is conducted over a wide range of normal stresses.

4.   Irrespective of the difference in the measured shear strength results between the two gripping systems, the findings demonstrated that the NP is the suitable gripping system for the geotextile-cushion (GTX-A)/geomembrane (GMB) interface and GMB/geosynthetic clay liner interface (GCL). The SP, on the other hand, is appropriate for GCL/Geotextile-filter (GTX-B) and the GTX-A/GMB/GCL/GTX-B interface.

Based on the results, it can, therefore, be hypothesized that the NP effectively grips when testing the GTX-A/GMB and GMB/GCL interfaces. However, it is important to note that the height of the nail plate teeth should not be greater than the thickness of the tested geosynthetics. Otherwise, the SP should be employed in the experiment.

5.   The use of appropriate gripping systems in geosynthetic interface shear strength will increase the level of accuracy, reproducibility and ease the interpretation of the shear strength test results. Consequently, safer, cost-effective and innovate designs of projects that involve the use of geosynthetic would be implemented.

6.   The determined shear strength characteristics using a single interface test approach in a large direct shear test, are higher as compared to the ones when a multi-interface test configuration is utilized, regardless of the gripping system and normal stresses considered. Therefore, multi-interface tests yield a conservative estimate of the shear strength for the tested geosynthetics.

6.3   Recommendations

This study involved experiments conducted on a 305 mm x 305 mm direct shear apparatus using three types of geosynthetics. Additional tests should be performed using different recommended direct shear devices i.e. 1.0 m x 1.0 m and the findings be compared with respective results obtained in this study. Also, there is potential for further study that can be addressed through extended investigations to either update existing literature or to provide new data for analysis. Therefore, the following recommendations are made for future studies:

1.   The research revealed the effects of gripping systems in the measured shear strength results between the NP and SP. Further investigation in quantifying the influence of these gripping system is required to refine the correlation generated between the results.

2.   The use of instrumentation that can capture images for the true engagement of the test specimens and the gripping system during the tests, is suggested. This is because the post visual examination of the failed samples cannot display a microscopic scale behaviour involved between the gripping system and test specimens.

3.   All commonly used gripping systems in interface tests should be investigated to reflect on which gripping system would be appropriate for various types of geosynthetic/geosynthetic interface tests. This system should avoid tensioning, transfer the applied shear stress across all failure surfaces, allow peak shear strength to occur anywhere and replicate the actual behaviour of the shear stress versus horizontal displacement of the tested specimens.

REFERENCES

AKS, 2018. *DT, HDPE Properties,*

Allen, J.M. & Fox, P.J., 2007. Pyramid-Tooth Gripping Surface for GCL Shear Testing. *Geosynthetics 2007 Conference Proceedings*, pp.1–9.

ASTM D4439, 2018. *Standard Terminology for Geosynthetics,*

ASTM D4439, 2006. Standard Terminology for Geosynthetics.

ASTM D5321, 2013. Standard Test Method for Determining the Shear Strength of Soil-Geosynthetic and Geosynthetic-Geosynthetic Interfaces by Direct Shear. *ASTM International*, pp.1–11.

ASTM D5321, 2017. Standard Test Method for Determining the Shear Strength of Soil-Geosynthetic and Geosynthetic-Geosynthetic Interfaces by Direct Shear 1. , pp.1–11.

ASTM D6243, 2018. Standard Test Method for Determining the Internal and Interface Shear Strength of Geosynthetic Clay Liner by the Direct Shear Method 1. , pp.1–12.

Baba, K. et al., 2012. Slope Stability Evaluations by Limit Equilibrium and Finite Element Methods Applied to a Railway in the Moroccan Rif. *Open Journal of Civil Engineering*, 02(01), pp.27–32.

Bacas et al., 2015. Frictional behaviour of three critical geosynthetic interfaces. *Geosynthetics International*, (5), pp.1–11.

Bacas, B. et al., 2015. Shear strength behavior of geotextile/geomembrane interfaces. *Journal of Rock Mechanics and Geotechnical Engineering*, 7(6), pp.638–645.

Bacas, B.M. et al., 2011. A new constitutive model for textured geomembrane/geotextile interfaces. *Geotextiles and Geomembranes*, 29, pp.137–148.

Bhatia, S.K. & Kasturi, G., *Comparrison of PVC and HDPE Geomembranes (Interface Friction Performance),*

Blight, G., 2008. Slope failures in municipal solid waste dumps and landfills: A review. *Waste Management and Research*, 26(5), pp.448–463.

Bouazza, A. et al., 2002. Geosynthetics in Waste Containment Facilities: Recent Advances. *7th International Conference on Geosynthetics*, pp.445–507.

Buthelezi, S., 2017. *Comparison of shear strength properties of textured polyethylene geomembrane interfaces in landfill liner systems.* University of Cape Town.

CETCO, 2014. Evaluating GCL Chemical Compatibility. *North America.*

Cilliers, C., 2018a. *Geosynthetics Engineering Course - University of Cape Town,*

Cilliers, C., 2018b. *Personal Conversation,*

Cilliers, C., 2018c. *Strongest and Weakest Interface on the Shear stress versus Normal stress*

grapg,

Covemaeker, J., 2017. Percentage change calculator - Calculate percent change.

Crystal Clear Maths, 2013. *Understanding Negative Leading Coefficients (Polynomials),*

David, J.S.L. & Robert, T.T., 1991. Geomembrane interface strength tests.

Department of Environmental Affairs (DEA) - South Africa, 2013. *Waste Classification and Management Regulations and Supporting Norms & Standards,*

Department of Water Affairs and Forestry, 1996. *South African Water Quality Guidelines,*

Department of Water Affairs and Forestry (DWAF), 1998. *Minimum Requirements for Wa Ste Disposal By Landfill,*

Dookhi, A.S., 2013. *The Lining of Steep Landfill Slopes in South Africa and the Applicability of the "Minimum Requirements for Waste Disposal by Landfill" by the Department of Water Affairs and Forestry.*

Eid, H.. et al., 1999. Effect of Shear Displacement Rate on internal Shear Strength of a Reinforced Geosynthetic Clay Liner. *Geosynthetics International*, 6(3), pp.219–239.

Eid, H.T. & Stark, T.D., 1997. Shear behavior of an unreinforced geosynthetic clay liner. *Geosynthetics International*, 4(6), pp.645–659.

Eid, H.T. & Stark, T.D., 1996. Shear Behavior of Reinforced Geosynthetic Clay Liners. *Geosynthetics International*, 3(6), pp.771–786.

Ernest, B., 2014. What percent error is too high? | Socratic. , pp.1–2.

Esterhuizen, J.J.B. et al., 2001. Constitutive Behavior of Geosynthetic Interfaces. *Journal of Geotechnical and Geoenvironmental Engineering*, 127(10), pp.834–840.

Fibertex, 2017. *Fibertex Geotextiles,*

Fox, J.P. & Ross, J.D., 2011. Relationship between NP GCL Internal and HDPE GMX/NP GCL Interface Shear Strengths. *Journal of Geotechnical and Geoenvironmental Engineering*, 137(8), pp.743–753.

Fox, P.J. et al., 1997. Design and Evaluation of a Large Direct Shear Machine for Geosynthetic Clay Liners.

Fox, P.J., 2010. Internal and Interface Shear Strengths of Geosynthetic Clay Liners. *3rd International Symposium on Geosynthetic Clay Liners*, pp.1–17.

Fox, P.J. et al., 1998. Internal Shear Strength of Three Geosynthetic Clay Liners. *Geotechnical and Geoenvironmental Engineering*, 124(10), pp.933–944.

Fox, P.J. et al., 2004. Laboratory Measurement of GCL Shear Strength.

Fox, P.J. & Kim, R.H., 2008. Effect of Progressive Failure on Measured Shear Strength of Geomembrane/GCL Interface. *Geotechnical and Geoenvironmental Engineering*, 134(4), pp.459–469.

Fox, P.J. & Stark, T.., 2004. State-of-the-art report: GCL shear strength and its measurement. *Geosynthetics International*, 11(3), pp.141–175.

Fox, P.J. & Stark, T.D., 2015. State-of-the-art report: GCL shear strength and its measurement – ten-year update. *Geosynthetics International*, 22(1), pp.3–47.

Geocomp, 2018. *Large ShearTrac I I I -Technical Specifications*,

Geosynthetic Institute (GSI), 2013. *Test Methods and Properties for Nonwoven Geotextiles Used as Protection (or Cushioning) Materials"*,

Geosynthetic Materials Association, 2016. *Geomembranes and Geosynthetic Clay Liners (GCLs)*,

Geosynthetic World, 2018. Layers of a landfill. , p.1.

Geosynthetics Interest Group of South Africa, 2018. *Three main engineering elements of the landfill*,

Gilbert, R.B. & Byrne, R.J., 1996. Strain-Softening Behavior of Waste Containment System Interfaces. *Geosynthetics International*, 3(2), pp.181–203.

Giroud, J.P., Darrasse, J. & Bachus, R.C., 1993. Hyperbolic expression for soil-geosynthetic or geosynthetic-geosynthetic interface shear strength. *Geotextiles and Geomembranes*, 12(3), pp.275–286.

Griffiths, D. V & Lane, P.A., 2001. Slope stability analysis by finite elements Slope stability analysis by ®nite elements. *Geotechnical*, 49(3), pp.387–403.

GSE Environment, 2017. Geomembranes for Landfill & Waste Containment.

Hammah, R.E. et al., 2009. Probabilistic Slope Analysis with the Finite Element Method. *American rock Mechanics Association*, 09, pp.1–19.

Hardie, P., 2018a. *Geosynthetics Engineering Course - University of Cape Town*,

Hardie, P., 2018b. Personal Conversation.

Hegde, A. & Roy, R., 2018. A Comparative Numerical Study on Soil–Geosynthetic Interactions Using Large Scale Direct Shear Test and Pullout Test. *International Journal of Geosynthetics and Ground Engineering*, 4(1), p.2.

James, G., 2018a. *Geosynthetics Engineering Course - University of Cape Town*,

James, G., 2018b. Personal Conversation.

Jayasuriya, A., 2017. The scientific definition of strain softening.

Jogi, M., 2005. A method for measuring smooth geomembrane/soil interface shear behaviour under unsaturated conditions. , (November).

Jones, D.R. V & Dixon, N., 1998. Shear strength properties of geomembrane/geotextile interfaces. *Geotextiles and Geomembranes*, 16(1), pp.45–71.

Juvinal, R.C. & Marshek, K., 1991. *Fundamentals of machine component design*,

Kalumba, D., 1998. *Kalumba Effect of Grading and Grain Size on the Friction Characteristics of a Sand Geotextile Interface.*

Kalumba, D., 2018. *Personal Conversation,*

Kavazanjian, E. et al., 2012. Performance based design for seismic design of geosynthetics-lined waste containment systems. In *Geotechnical, Geological and Earthquake Engineering.* pp. 363–385.

Kaytech Engineered Fabrics Ltd, 2013. *GCL - X 800 900 10,*

Kaytech Engineered Fabrics Ltd, 2015. *Geotextile - Bidim A 10,*

Koerner, R.M. & Koerner, G.R., 2007. *Interpretation(s) of Laboratory Generated Interface Shear Strength Data for Geosynthetic Materials With Emphasis on the Adhesion Value,*

Koerner, R.M. & Soong, T.-Y., 2005. Analysis and design of veneer cover soils. *Geosynthetics International,* 12(1), pp.28–49.

Koerner, R.M. & Soong, T.-Y., 2000. Stability Assessment of Ten Large Landfill Failures. *Geotechnical Special Publication,* 103(103), pp.1–38.

Lin, H. et al., 2014. An improved simple shear apparatus for GCL internal and interface stress-displacement measurements. *Environmental Earth Sciences,* 71(8), pp.3761–3771.

Lopez-Anido, R.A. & Naik, T.R., 2000. *Emerging Materials for Civil Infrastructure: State of the Art - Google Books* G. T. Fry, D. A. Lange, & V. M. Karbhari, eds., ASCE.

McCartney, J. & Swan, R.H., 2002. Internal and Interface Shear Strength of Geosynthetic Clay Liners (GCLs): Additional Data by. , (June), pp.1–36.

McCartney, J.S. et al., 2009. Analysis of a Large Database of GCL-Geomembrane Interface Shear Strength Results. *Journal of Geotechnical and Geoenvironmental Engineering,* 135(2), pp.209–223.

Merrill, K. & O'Brien, A., 1997. Strength and Conformance Testing of a GCL Used in a Solid Waste Landfill Lining System. *American Society for Testing and Materials,.*

Minitab Inc, 2018. Regression Analysis: How Do I Interpret R-squared and Assess the Goodness-of-Fit?

Mitchell, J.K. et al., 1993. The Kettleman Hills Landfill Failure: A Retrospective View of the Failure Investigations and Lessons Learned. *International Conference on Case Histories in Geotechnical Engineering,* pp.1–15.

Miuzzi, M., 2012. *Inclined Plane Tests: Determination of Friction on Geosynthetic Interfaces,*

NILEX Civil Environmental Group, 2010. *Geosynthetic Clay Liners (GCL's),*

Ongodia, J.E., 2017. *Geotechnical Engineering Design of a Tunnel Support System.*

Oriokot, J., 2018a. *Geosynthetics Engineering Course - University of Cape Town,*

Oriokot, J., 2018b. Personal Conversation.

Qian, X., 2008. Critical interfaces in geosynthetic multilayer liner system of a landfill. *Water Science and Engineering*, 1(4), pp.22–35.

Qian, X. et al., 2003. Translational Failure Analysis of Landfills. *Journal of Geotechnical and Geoenvironmental Engineering*, 129(6), pp.506–519.

Qian, X. & Koerner, R.M., 2004. Effect of Apparent Cohesion on Translational Failure Analyses of Landfills. *Managing*, 130(1), pp.71–80.

Rabie, M., 2014. Comparison study between traditional and finite element methods for slopes under heavy rainfall. *HBRC Journal*, 10, pp.160–168.

Ross, J.D., Fox, P.J. & Olsta, J.T., 2010. Dynamic shear testing of a geomembrane/geosynthetic clay liner interface. In *9th International Conference on Geosynthetics*.

Rouncivell, W., 2005. *Experimental Investigation of the Shear Strength Characteristics of A Geosynthetic Clay Liner and its Application In A Local Landfill Lining System*. University of Cape Town, South Africa.

Ruan, X.-B. et al., 2013. Effect of the amplification factor on seismic stability of expanded municipal solid waste landfills using the pseudo-dynamic method #. *J Zhejiang Univ-Sci A (Appl Phys & Eng)*, 14(10), pp.731–738.

Russell, D. et al., 1998. Shear strength properties of geomembrane/geotextile interfaces. *Geotextiles and Geomembranes*, 16, pp.45–71.

Samirsinh, P., 2016. Functions and selection of geosynthetics.

Sarsby, R.W., 2007. *Geosynthetics In Civil Engineering*,

Scheirs, J., 2009. *A guide to polymeric geomembranes*, Wiley.

Shukla, S.K., 2012. Geosynthetic Engineering – Basic Concepts.

Sikwanda, C. et al., 2018. Review of Effects of Geosynthetic Gripping System in Interface Testing. *Springer*, pp.1–10.

Soleimanian, M.R. et al., 2016. *A Novel Direct Shear Apparatus to Evaluate Internal Shear Strength of Geosynthetic Clay Liners for Mining Applications*.

Stark et al., 2011. Comparison of Single and Multi Geosynthetic and Soil Interface Tests. *Geosynthetics International Journal*, 1(404), pp.2–55.

Stark, T.. et al., 1998. Unreinforced Geosynthetic Clay Liner Case History. *Geosynthetics International Journal*, 5(5), pp.521–544.

Stark, T.. & Choi, H., 2004. Peak vs. Residual Interface Strengths for Landfill Liner and Cover Design. *Geosynthetics International*, 2(6), pp.1–7.

Stark, T.D., 2018. *Slope Stability Concepts and Failure*,

Stark, T.D. et al., 2015. Strength Envelopes from Single and Multi Geosynthetic Interface Tests. *Geotechnical and Geological Engineering*, 33(5), pp.1351–1367.

Stripp, D., 2018. Personal Conversation.

Suntech Geotextile, 2018. Functions of Geotextiles.

Swan, R. et al., 1997. Short-Term and Creep Shear Characteristics of a Needlepunched Thermally Locked Geosynthetic Clay Liner. *American Society for Testing and Materials*, pp.97–110.

Textile Centre of Excellence, 2018. Geotextiles Overview | Geotextiles | Market Sectors | Knowledge | Huddersfield Textiles.

Thiel, R.S. & Criley, K., 2005. Hydraulic Conductivity of partially prehydrated GCLs under high effective confining stresses for three real leachates. In *Proceedings of the Geo-Frontiers Conference (Austin)*.

Trauger, R.J. et al., 1997. Long-term Shear Strength Behavior of a Needlepunehed Geosynthetie Clay Liner. *American Society for Testing and Materials,*, pp.103–120.

Triplett, J.E. & Fox, J.P., 2001. Shear Strength of HDPE Geomembrane/Geosynthetic Clay. *Geotechnical and Geoenvironmental Engineering*, 127(6, June), pp.543–552.

United States Dept. of Labor, 2012. Your Step Matter.

Veylon, G. et al., 2016. Performance of geotextile filters after 18 years' service in drainage trenches. *Geotextiles and Geomembranes*, 44, pp.515–533.

Visser, W., 2018. *Critical Interface In Landfills*, Cape Town, South Africa.

Westlake, K., 1995. *Landfill waste pollution and control*, Albion Pub.

WordPress, 2010. Advantages and Disadvantages Of Commonly Used Synthetic Geomembranes | Ieccovers's Blog. *Blog at WordPress.com*, p.1.

Zanzinger, H. and & Alexiew, N., 2000. Prediction of long term shear strength of geosynthetic clay liners with shear creep tests.

Zanzinger, H. & Saathoff, F., 2010. Shear creep rupture behaviour of a stitch-bonded clay geosynthetic barrier, 3rd International Symposium on Geosynthetic Clay Liners.

Zelic, B.K. et al., 1998. Shear strength testing on a GCL. , pp.1–6.

Zornberg, J.G. et al., 2005. Analysis of a Large Database of GCL Internal Shear Strength Results. *Journal of Geotechnical and Geoenvironmental Engineering*, 131(3), pp.367–380.

Zornberg, J.G. & Christopher, B.R., 1999. *Geosynthetics*,